I0475332

# The Project Book

## A Simple and Direct Approch to Project Management

**Abdulla J. Alkuwaiti**

M.Sc. in Project Management, PMP, RMP, MSP, PRINCE2

# Table of Contents

# Table of Contents

To Fakhera, Aysha, Mariam, Ahmed and Mohamed

# About the Author

Abdulla Al-Kuwaiti finished his B.Sc. in Systems Engineering at the University of Arizona in 2000. Shortly after, he joined a major oil company as a Safety Engineer. Since then, he has been introduced to project management, trained in it and is practicing it on a daily basis. Abdulla is a believer in project management and constantly looks for the means to simplify it so that its principles and tools may be used more often. He achieved his Masters in Project Management in 2007 and now works as a Program Manager. Abdulla is RMP- and PMP-certified by the Project Management Institute in the USA. He is also a PRINCE2 and MSP-certified by the Association of Project Managers in the UK. His interests are in risk management, program/ project management and benefit management.

**You may contact the author at: alk.books@gmail.com.**

**www.kuwaitat.net**

## Author's Statement

In 2010, I conducted a training course on project management. My goal was to simplify project management and present it in an attractive way since I noticed that many project managers don't use project management tools and methodologies. In that training I gave particular attention to the importance of the planning phase of the project and to how ignoring and bypassing it can increase the chances of project failure.

It has become very easy to get information about project management nowadays: Just type "project management" into any reliable search engine and you will get links to thousands of articles and much information about the subject. Yet, regardless of this wealth of information and knowledge, there is a lack of implementation. Many projects, large and small, are undertaken without implementing project management methodologies, and I believe a major reason for that is the absence of a simplified framework that takes the hand of the project manager and guides him/her STEP BY STEP throughout his/her project. This is what I have tried to put forth in "The Project Book."

I hope this book will assist project managers in managing their projects, especially during the planning phase. I would suggest that a ring-binder be dedicated to every project as a place where templates for each element of the project can be placed and filled out in a team setting.

In addition, I have written an article about my training courses and the techniques I used. The article was published in the virtual library of the Project Management Institute and can also be found at the end of this book.

# Chapter One
# Introduction to Project Management

Our life, professional and personal, is filled with projects. Projects are the means by which we achieve our goals, and the mere naming of a group of tasks as a "project" will help us think of the best ways to accomplish them while considering available resources like time and money. As such, we can define a project as: a group of activities through which we achieve some goals in a specified period of time.

A project has special characteristics, such as:

- Has goals to achieve: There are clear reasons why we undertake a project;

- Has a product: When the project ends, it must deliver something (for example, a physical product, a document or a service);

- Has a specific time period: It has a beginning and an end. Even if the project lasts for years, it must end on a specific date, which is the day it delivers its intended product/s. As such, routine tasks (for example, customer service activities) are not projects because they don't have a specified or expected end date;

- Is unique: Each project is different. Even projects that produce similar products will be different in their inner and outer environmental conditions. This uniqueness demands that the project manager treat every project as a set of special and distinctive activities. (No "copy and paste" here.);

- Requires the use of particular resources: Examples are labor, money and even mental resources for thinking and brainstorming;

- Involves a large amount of uncertainty: Unlike routine tasks in which you know exactly what to do, a project may unfold in as many different ways as you can imagine.

## Examples of Projects

- Construction projects, such as building towers, tunnels, gardens;

- Personal projects, such asstarting to do your master's degree or preparing for your wedding;

- Research projects, such as conducting a study on the native people in the Amazon;

- IT projects, such as creating a database for the employees in your company;

- Event-related projects, such as planning exhibitions and parties.

## Definition of Project Management

We have defined a project, but what about project management? The key word in defining project management is APPLICATION. Project management is the process of applying the methodologies and practices that are well accepted in the field of project management for the purpose of increasing the possibilities of project success and achievement.

Various scholars and professionals have observed different projects and studied the elements that caused them to succeed or fail. They used their observations to organize and put forth methodologies and tools to help project managers improve in managing their projects. For instance, experts found that in order to have a successful project, we must apply the principles of time management, risk management and cost management.

Please note that project management is an established science which is independent from the technical field of the project. To illustrate this point, consider the project of constructing a residential building. The technical aspect of this project is "construction," in which you will use fields of science such as civil engineering, architecture and structural engineering. However, in addition to the technical fields, you will use project management as a dominant science; imagine it as an umbrella under which you will practice or apply the technical knowledge and expertise needed by the project. Now, my argument based on the above discussion is that if becoming a civil engineer requires formal education, why shouldn't it be the same for project management, which has many fundamentals and methodologies that need to be understood and studied? I have seen many people bearing the title of "project manager" without any formal knowledge of the science of project management. I am not saying that you can't have good, or even excellent, project managers who excel in their work as a result of their talent and experience rather than formal training, but such are the exception. Moreover, you don't have to spend years in college studying project management before you manage projects (though that could prove beneficial, of course). The message I want to share through this book is for project managers to look at project management as a science; they need to get to know its fundamentals so that when they build upon their experience, they will have the best of both worlds: the theoretical and the practical. In this book, I have tried to present these fundamentals in an easy and practical way.

## Different Schools of Project Management

The science of project management is not absolute and firm; rather, it allows space for disagreement, discussion and different opinions. This is rather beneficial to project managers as it makes project management flexible; there might be more than one solution to a problem or more than one tool that can be used to resolve a certain situation.

Due to the flexibility of project management, different schools have emerged, presenting project management from varying points of view. For example:

- Project Management Institute (PMI);

- Association for Project Management (APM);

- International Project Management Association (IPMA);

- The Australian Institute of Project Management (AIPM);

- Project Management Association of Japan (PMAJ).

**Project Management vs. Everyday Management**

In addition to having tasks that are combined into projects, organizations also have day-to-day tasks to accomplish, such as customer service, routine maintenance and internal/external communication. For such recurring tasks, an organization seeks harmony and the repetition of the same level of performance. The best way to achieve this is through normal everyday management, with set policies and procedures for employees to follow in order to produce the required results every time.

Projects, on the other hand, differ from routine work in that they are unique and the degree of uncertainty about how they will unfold over time is much higher. The principles of project management are best suited for tackling these differences because they:

- Take into consideration the uniqueness of every project through the use of an execution plan tailored for each project;

- Offer different tools and techniques for planning project activities, which can decrease the degree of uncertainty;

- Encourage creativity in finding the best use for the available resources.

Consider the example of a commercial bank which has both projects and routine work. An example of routine work would be responding to customers at the counters, a task for which the bank will want to train its employees to follow (and repeat) certain procedures, such as the steps for opening a checking account for a new customer. On the other hand, the bank might have a project such as creating a website to provide E-services for its customers. The bank can not simply use routine procedures to create the site because it is a unique task happening for the first time and there is no specific information yet as to how it will be accomplished (the project might face unanticipated requirements and risks). As a result, it is better for the bank to apply project management, which consists of many tools, such as scheduling and risk management, that will help deliver the website.

## Project Life Cycle

By observing different projects, many professionals and experts conclude that projects have four phases that they pass through as follows:

- **Project Initiation:** where the idea of the project is investigated;

- **Project Planning:** involves describing the idea for the project, determining the required resources and setting a time frame in which to execute it;

- **Project Execution:** where the project plan is implemented to deliver the project objectives;

- **Project Closure:** involves the official closeout of the project and the recording of lessons learned in the process.

It is worth mentioning that dividing projects into these phases is beneficial as it will help you as a project manager to distribute your effort across these phases and will give you a general road map to follow throughout the project.

### How a Project Progresses in Its Life Cycle

First, from the point of view of resource usage, during the initial stages of the project, minimum resources are needed, but as the project progresses in its life cycle, more and more resources are used, especially during the execution stage. This information is very important as you can infer from it that it is cheaper to initiate changes to a project during its initial days when fewer resources have been spent. The following diagram illustrates this idea:

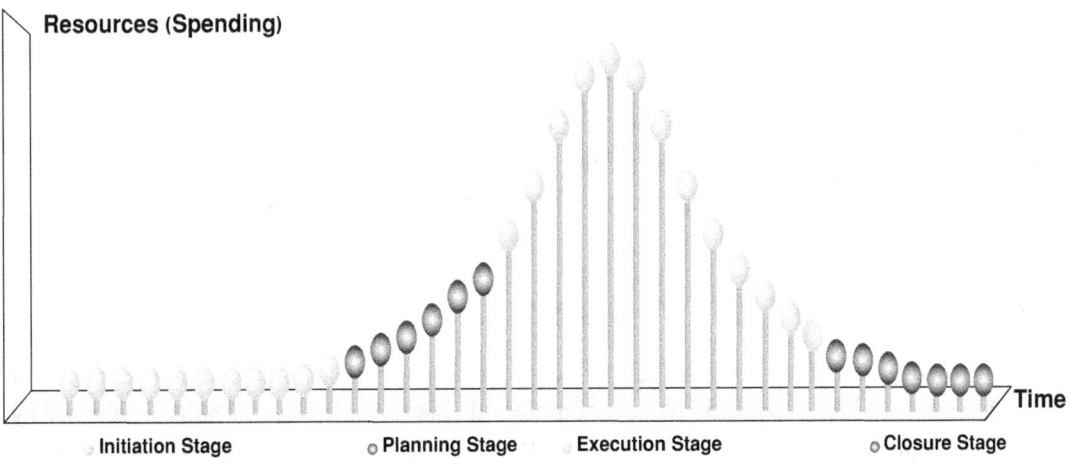

Second, from the point of view of levels of knowledge needed, during the initial stages of the project, there is a lot of ambiguity and lack of knowledge regarding how the project will actually progress (simply because we are not certain of the future). However, with the progress of the project, things start to become clearer and we can discern whether the actual project execution is going in the direction predicted in the project plan. The following figure shows the level of uncertainty throughout the project life cycle:

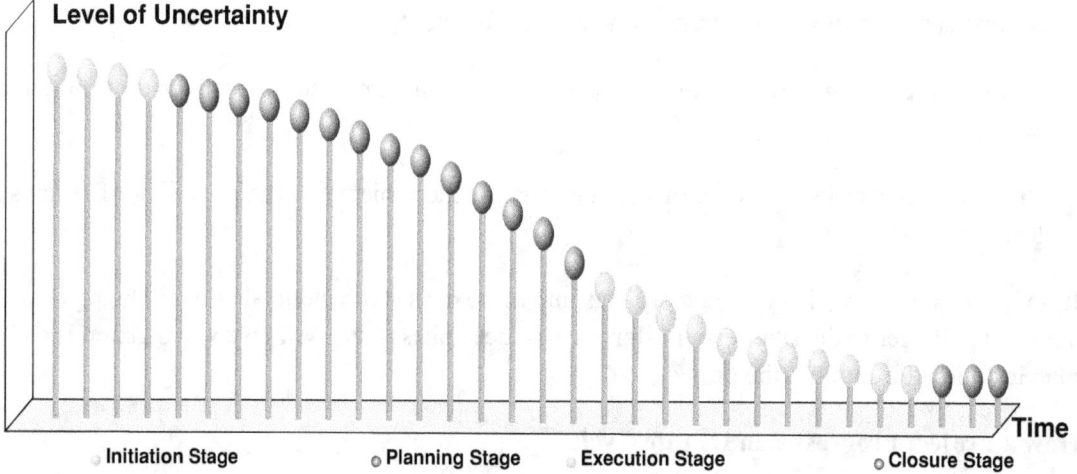

Knowing that the level of uncertainty is highest at the beginning of the project should lead project managers to exercise different methods and to use as many tools and techniques as possible to ensure the development of clarity and precision. As will be shown in this book, the use of different templates, especially the risk template, will help greatly in reducing the level of uncertainty.

**The Design of the Project**

In many projects, there is a stage dedicated to the design of the product, whether it be a building, a garden or an IT system. Some project managers might think this design is part of the project planning phase, but it is actually part of the project execution because you need to plan for the design first during the project planning phase, and then undertake the actual design in the project execution phase.

**The Importance of the Project Planning Stage**

There is no disagreement between the different schools of project management that project planning is essential for successful project completion. Nevertheless, you might be amazed by the number of projects that are executed with little or no planning at all. Reasons for bypassing project planning phases might be:

## 1 Difficulty of Planning:

- Project planning has many different elements, and covering them will require a lot of effort.

- Often project managers lack a simplified framework for how to start and complete a project plan.

- Project planning requires training in the different elements of a project. Such training might not be available for some project managers.

## 2 The Benefit of Planning Might Not Be Clear to Some Project Managers:

- The benefit of planning might not be clear because the plan is not linked in a realistic way to project execution. For example, during the risk management that takes place during project planning, many risks can be identified and measures to control them can be suggested and incorporated into the plan. But if these control measures are not implemented, then the whole exercise of that element of planning (that is, risk management) will just go to waste.

- Some project managers (especially fresh graduates) might give you the following argument: "The efforts of the planning phase are just a waste of time; planning is a way of predicting the future, and in many cases the future doesn't happen as expected, so the plan turns out useless." They might add, "Wouldn't it be better to just start executing the project and make adjustments on a day-to-day basis, responding to the events as they pop up?" It might be lack of experience and ignorance of the benefits of planning that make some project managers present such an argument. The best way to reply to it is by revealing the benefits of planning and giving examples.

## 3 Fear of Change:

- If project managers get used to bypassing project planning for a long time, they will most likely resist the idea of making an official and documented plan. This resistance might emerge from the fear of the extra work that is required for planning or the fear of monitoring (since without a project plan, progress cannot be monitored in a precise manner).

In this book, I will try to present a way to simplify planning by introducing templates for the different planning elements in a structured and easy-to-follow framework. In addition, I will give examples of how planning can be reflected in reality during project execution.

## Project Management: Art and Science

I want to make it clear that project management depends on creativity as much as it depends on science. The scientific part of project management tells you "what" you have to do and the artistic side tells you "how" to do what you have to.

## Project Management as a Science

My focus in this book is on the scientific side of project management. This side provides you with the methodologies, tools and techniques to follow for the best chance of having a successful project. I decided to focus on this side because it is the foundation and starting point every project manager must go through, and into which experience and intuition can then be integrated. Also, project managers can learn about the methods of project management through books and training courses, but the creativity part depends more on talent and experience. Examples of how I have focused on the scientific part of project management in this book are the provision of a template for each of the project management elements and a simplified brief on the theory behind each element. There are also many tools and techniques presented in this book.

## Project Management as an Art

By "art," I am referring to the use of creativity on the part of the project manager, which involves ingenuity and seeing things from a new prospective. There is no doubt about it, a project manager needs a good deal of creativity. We have defined a project as a unique work; that uniqueness will certainly demand a unique and creative management approach. If projects were to be accomplished through merely "copying and pasting," then we might be settling for a "robotic" style of management, but in reality a project manager will be faced with many unexpected and special situations that require creative (and immediate) decisions. Learning the artistic side of project management is much harder than learning the scientific side because it is not a definitive science; rather, it depends on complex and interrelated elements, such as experience and "soft skills," such as communication and time management. Here are some examples of the creative side of projects:

- Leading the project team in delivering the project objectives;
- Solving disputes and conflicts;
- Decision-making;
- Skills for running meetings;
- Being diplomatic with the different stakeholders and trying to please them all without sacrificing the project objectives.

From the above examples, we notice that there is not just one rule that applies to all of them. For example, to lead the project team, the project manager might find many theories on different leadership styles, but selecting the right style will depend on the creativity of the project manager and his/her ability to evaluate the situation and respond to it.

The artistic side of the project manager will develop with the variety of projects that he/she participates in; however, the project manager must first be keen on learning sound theories and methodologies in project management upon which the acquired experience can be built.

**When There Is More than One Project Manager for One Project**

Some projects will have more than one project manager. Consider a governmental project to construct a  neighborhood park.  Here you may need three project managers, one from the government, one from the consultant company that will design the park and one from the contracting company that will actually undertake the construction.  In this case, there should be no conflicts among their roles; rather, the project manager will "manage" the project manager from the consultant company who, in turn, will "manage" the project manager from the construction company.  I present this situation because it is rather common and often the project manager from the government (or sponsoring organization) leaves the burden of project planning on the consulting company, which is ill-advised as they are not expected to be the most concerned party in ensuring project success as is the sponsoring company's project manager. In this situation, the project plan will be created by the project manager from the sponsoring company, updated  and completed by the consulting company and reviewed by the contracting company.  The following figure illustrates the hierarchy of responsibilities in such a case:

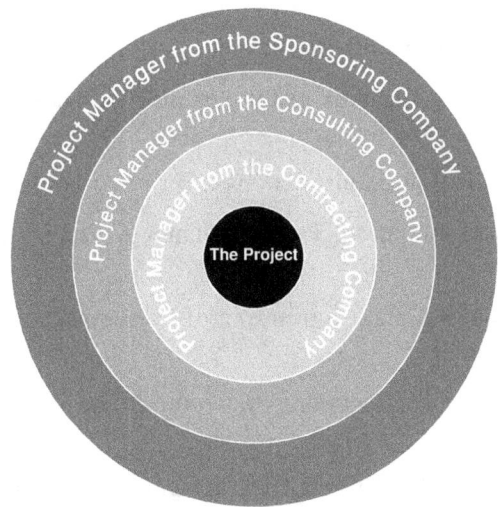

**The Importance of Documentation**

Often I hear from fresh graduates assigned to projects who feel that filling out project templates is just a writing exercise with no added benefits (aside from wasting their time). I tend to sympathize with them as many project management templates are becoming very lengthy, as if they were written by lawyers who want to present every single detail no matter how trivial it is. Also, it is my opinion that many templates do not capture the essence for which they are created, which is to be interrelated with the other project templates and to actually be used during project execution. I hope that this book recaptures the role of templates.

But first, let us go back to the argument presented at the beginning of this paragraph, as there is much to be said about it. Some project managers do question the need to document their planning in writing, just as many fresh graduates question the value of documentation, as they just want to "jump on the project and execute it"; documentation is considered a "desk job" and not "real fieldwork."

Yes, documentation (done either in templates or by other means) will take time, but it is nothing to be taken lightly. When you document, you leave physical record for others, enabling them to review what you wrote, improve on it and use it further. Documentation will help you build and progress, as development is not built on ideas that are still in a person's mind but is accomplished when those ideas are written down and presented to others. Additionally, the need to document is becoming an internationally accepted wisdom as ALL international associations require documentation as an important element for those interested in joining them (for example, the ISO organization). Finally, when you write down the different planning elements of your project, you will actually improve on your ideas. Think about it: when you transfer an idea from your head to a piece of paper, your mind makes some mental calculations, assessing the idea and tweaking it to make sure it is worthy of being written down (since your mind knows that once your idea is written down, others can review it, and though your mind might trick you into thinking you have a good idea, it is certainly not foolish enough to make a record of it).

**How Might Project Managers Use This Book?**

If you will be managing a project (for work or for personal purposes), you can start using this book right away, simply by doing the following:

1 Buy a ring binder and divide it into four sections.

2 Label the sections: Project Initiation, Project Planning, Project Execution and Project Closure.

3 Visit **www.theprojectbook.net**, make a print-out of the templates and arrange them in the different sections of the binder.

4 Read the section of this book concerning the project feasibility study and fill in the corresponding template.

5 Go through all the sections one by one and fill in the templates as you progress through your project.

Ultimately, you will create a full folder that will not only document your project information but will include all your (and your team's) ideas and plans for a successful project.

**How Might an Organization Use This Book?**

Using the "Project Book" method in a company will depend on its maturity in managing projects and its existing methodologies and techniques. This book is mainly geared toward companies that are in the beginning to intermediate stage of maturity in applying project management to their projects, and is aimed at helping them establish a clear framework that can be applied to different projects to help execute them in the best possible manner. An organization can use this book in two ways:

**First:** As a training tool and a reference for its project management staff;

**Second:** As a gateway to audit the progress of projects as described in the following table:

| Gate | How to Use the Book |
|---|---|
| 1 Approval to start project planning | To give the go-ahead to start planning for a project, there must be a "Project Book" in the form of a ring binder with the first section (that is, Initiation) filled out and endorsed by senior management. |
| 2 Approval to start project execution | Budget will not be assigned to projects and they will not be started unless the section about project planning is completed with all the templates, and reviewed and endorsed by top management. |
| 3 Officially declaring a project completed and closed | A project will be considered closed when it has a complete project book |

**Note:** To implement the gateways, the organization will need to do some preparation work, such as:

1 Establish committees (for each project or for the whole company) that meet regularly to review the project book binder;

2 Declare the project book binder an official folder that is linked to different departments of the company (for example, the finance department will accept its templates to provide financial resources).

# Chapter Two
# The Project Book

The Project Book method is based on the idea of a template for each element of the different project stages. However, I would like to note that the aim of the book is not just to fill in templates as an isolated writing exercise, but rather to use the templates to achieve two things: First, templates will allow us to advance our project without missing any element of the project stages because they will provide a clear framework to follow. Second, in the process of filling out templates, sufficient thinking and analysis will be devoted to the unique requirements of the project at hand. To implement the idea of the Project Book, we need to fill out the templates as follows:

1  Each template is filled in one at a time, following the sequence in the Project Book framework.

2  Templates are filled out by a group of people and not just the project manager (whenever it is possible).

3  Brainstorming is required for filling in the required data in the templates.

4  Relevant documentation must be reviewed before filling out a template.

5  Interdependency between templates is to be considered. In many cases, information between templates will be exchanged, so care must be taken to ensure that there is no conflicting information between templates.

6  Filling out templates is a dynamic process in which, if new information emerges, it is absolutely possible to go back to a template and amend it as required.

When I designed the templates, I intentionally tried to put each one on only one page, as I believe there is a correlation between the size of the template and its simplicity. However, this doesn't mean that a template can be completed in a few minutes as its size might appear to suggest. On the contrary, it is expected that to fill out one template, many documents will be reviewed and healthy discussions will take place among the project team members, similar to an organic perfume for which thousands of flowers are needed to produce a very small amount.

The following model shows why I have a high regard for templates. Using templates will serve as the ignition source for many processes that will directly and indirectly contribute to project success.

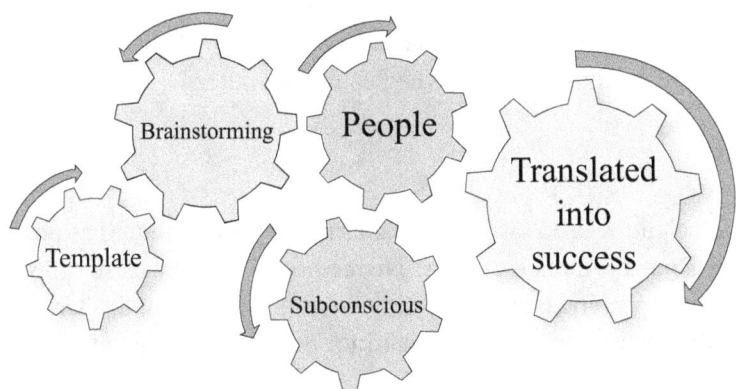

The Project Template Model

**Advantages of Using Templates**

- They reduce disorganization and provide a sense of control in managing the project.

- They provide a somewhat concrete sense of the different elements in the project stages by providing documentation for each element, thus making it easier to think about these elements rather than the elements remaining in the form of abstract ideas.

**Notes:**

- I mentioned that each template must be filled in separately. The reason is that you will often require information from previous templates in order to enter information into other ones. This will create interdependency between templates to give the project manager more confidence in them, encouraging him/her to use them.

- It might be impossible to cover all the required information for one project element in a single template, just as the differences between companies and projects will demand customized templates. For such, I recommend that after going through this book, you review the templates and make changes to them as you see appropriate. For example, you might want to customize the templates to include the terminology used in your company (for example, some companies might use the title of project director or coordinator instead of project manager).

- When designing the templates, my concern was not to make them look attractive and artistic as much as to make them practical and to include all the required information.

- A major threat to filling out templates is the use of the "copy-paste" method. If a project manager just copies information from the templates of similar projects, then the essence of the templates is lost. Also, the uniqueness of each project will not be addressed properly, and templates will just be pieces of paper isolated from the real execution of the project. I, therefore, strongly recommend that templates be filled out by hand, as such will discourage the copying and pasting that is so easy when using computers.

**Brainstorming**

Brainstorming is a technique used for generating as many different ideas as possible to solve a particular problem. Brainstorming is an effective method in which we use simple techniques to enhance thinking. Examples include: verbalizing what is in your mind, sharing your thoughts with others and writing your ideas on a big board for others to see and add to them.

Using brainstorming when filling in the templates is one of the most important goals of the Project Book for the following reasons:

- Using brainstorming urges you to ask others to participate since you want to generate as many ideas as possible;

- Filling out templates through brainstorming reduces the chance of "copy-paste" as participants produce new and innovative ideas;

- Brainstorming helps in building the project team and introducing harmony as members work together and share ideas; it helps build trust and understanding.

**How to Get the Greatest Number of Ideas**

You can maximize the number of ideas produced during brainstorming by following these rules:

1  Invite a suitable number of interested participants (a good number is 3-7, but this depends on the situation, project needs and your judgment).

2  Act as a facilitator by leading the discussion, writing down the ideas and making sure that every participant has the chance to speak.

3  Clearly explain the problem or situation at hand in order to generate appropriate ideas.

4  Encourage everyone to participate.

5  Use a large writing board to record the ideas.

6  Focus on the quantity of ideas, not their quality (quality of ideas will be assessed later).

7  Use techniques such as grouping the generated ideas and building on them to generate new ones.  Also, if you notice that some attendees are not participating, you may ask them to share their ideas by writing them down on a piece of paper.

8  Use encouragement and avoid conflicts as much as possible.

**How to Fill In the Templates by Using Brainstorming**

Note that the project manager should act as a facilitator in these meetings by doing the following:

**First:** Preparing for the Meeting

1 Determine the reason for conducting the session (in this case, it will be to fill out a particular template).

2 Invite people whom you think are interested in the project and would most likely participate actively in discussing the project and generating ideas. You should also look for experts on the subject (for example, when filling out the cost template, you might invite people from the finance department). Make sure you invite an appropriate number of participants; I recommend inviting between 3 and 7 participants.

3 Prepare yourself for the meeting. For example, before you start the meeting, think of questions that may stimulate ideas from the participants.

4 Prepare logistically for the meeting. For example, book a meeting room and provide writing utensils.

**Second:** During the Meeting

1 Divide the meeting into two parts, first to capture ideas, and second to assess the ideas and finalize completion of the template.

2 Distribute copies of the template to be discussed to the participants, and explain the purpose of the brainstorming session (which is to fill out the template with the best ideas possible).

3 Start the discussion and encourage the flow of ideas; write them down.

4 Once a good number of ideas is generated, assess them with the group and select the best ones.

5 Fill out the template.

> The idea of brainstorming might seem very simple (it really is). Unfortunately, it is not widely practiced in project management, perhaps because it takes more time to generate ideas and make decisions with a group than to do individual work. Brainstorming also requires the project manager to be fairly democratic and accept the views of others. Brainstorming sessions should, therefore, be pre-planned and managed efficiently to prevent lengthy meetings that cause boredom among participants and feel like a waste of time.

**Examples in This Book**

In this book, you will find a complete example of how to prepare a Project Book binder. The sample illustrated in this book is for a municipal project, constructing a park in a residential neighborhood. There is also a secondary example on dinner planning to illustrate the possibility of using the Project Book method for personal projects in addition to corporate projects. However, please note that you may apply the Project Book method to any project, no matter how large or small, and in any field.

**Note**: In Chapter Eight you will have the opportunity to practice making project books as you will find two exercises, one about making a website, and one about planning a family vacation.

**The Framework of the Project Book**

I have mentioned the elements of the Project Book and the framework several times without explaining them. Our journey through the different elements will start from the next chapter as we go through each element and fill out each corresponding template. So, buckle up!

The following graphic illustrates the framework of the Project Book, including the different elements of a project:

| | |
|---|---|
| **1** **Project Initiation** | The birth of the idea |
| | Return-on-investment |
| | Feasibility of the idea |
| | The project charter |
| **2** **Project Planning** | The team charter |
| | Review of lessons learned |
| | Stakeholders managment |
| | Information communication plan |
| | Project scope |
| | Project map |
| | Quality plan |
| | Time plan |
| | Money and resources |
| | Contractors selection |
| | Change management & Progress reporting plans |
| | Risk management |
| **3** **Project Execution** | Implementing the templates and the Project Execution Board |
| | Project monitoring |
| **4** **Project Closure** | Lessons learned |
| | Project closure |

# Chapter Three
# Project Initiation

A project begins with a mere idea of achieving some goals. However, not all ideas are suitable or possible to turn into projects. Here, then, comes the role of the initiation stage: to distinguish suitable ideas from those that cannot be implemented for one reason or another. We may call the initiation stage the pre-project stage during which the following is done:

- Presenting the idea and its potential to be turned into a project;

- Showing that the project suggested by the idea has the potential to serve company objectives (short or long term);

- Securing official approval to plan for the project;

- Assigning a project manager.

To achieve the points mentioned above, we will use two templates: the feasibility study template and the project charter template. Usually, if the idea is accepted, a project manager is assigned to the project at the end of this stage; hence, at this point, the person who came up with the idea will be the one responsible for presenting it and filling out the required templates. In some cases, however, it might be possible to immediately identify a suitable project manager for the proposed project, and it would then be very beneficial to include him/her in filling out the templates at this stage.

**The Birth of a Project Idea**

The idea for any project, small or large, is born in the mind of an individual. Reasons for the emergence of such ideas might be:

- To satisfy organizational goals and objectives;

- For self- or company improvement;

- To respond to new laws;

- To respond to a challenge (such as proving a scientific theory);

- For adventure (such as climbing the tallest mountain).

**Example: Constructing a Community Park**

- The birth of the project idea

We will use the example of the park from now on. Let's assume that in the municipality of a particular city, there is a landscape engineer named Amanda working in the department of parks and recreation. Amanda sends an email to her boss suggesting that they construct a park on the northern side of the city. Amanda supports her idea with some facts, such as the absence of any parks in that area and the population of the neighborhoods surrounding the proposed location. In addition, Amanda mentions that building parks falls in line with the municipality's goals and objectives of providing green landscapes to serve the different communities within the city.

The manager of the parks department is pleased with the idea and tells Amanda that if the idea is accepted, he would make her the project manager. He instructs her to conduct a feasibility study for the park and present it during the monthly senior management meeting. He also instructs her to contact the urban planning department to make sure the proposed location is available.

## Project Feasibility Study

> From now on, you will see the following pattern in every section of this book:
>
> - First: The project management element will be explained.
> - Second: The relevant template for the project element will be filled out in relation to the example of constructing the community park. Filling out the template with the example should serve as a description of the template itself.
> - Third: The benefits gained from filling out the templates will be discussed.

New ideas are usually accompanied by a lot of enthusiasm, which might hinder the proper and thorough examination of their effectiveness and practicality. The feasibility study comes into play here, so that after the idea is presented, its usefulness is assessed by asking the following questions:

- Am I (or the organization) in real need of this project?

- Can I (or the organization) actually execute and complete the project?

The importance of conducting a feasibility study for a project is that it helps us in making the decision of "Go" or "No Go" since, once you start the project planning phase, it will consume a lot of time (and possibly resources). You had better be confident that the original project idea is worth pursuing.

### Assigning a Feasibility Study to a Specialized Company

A feasibility study should decrease the level of uncertainty and fogginess that accompanies the birth of most projects. Sometimes, however, an organization cannot reduce the uncertainty enough to encourage it to proceed with a project, especially in the case of a large and innovative project. As a result, the organization might want to hire a specialized company just to study the feasibility of the idea in question. This might cost extra money, but it will prevent two possible losses:

The first possible loss: That you continue a project you cannot complete successfully.

The second possible loss: That you decide not to continue a project that would prove beneficial.

The template we are going to use for assessing the feasibility of the idea is intended to give a strategic look at the project and how well it fits into the organizational goals. However, depending on the experience of the company with the intended project, you might find that after filling out the feasibility study template, there is a lot of missing information, in which case you might decide to hire a specialized company to make a feasibility study. In such a case, I would advise you to dedicate a separate Project Book binder to it and to treat the study as a separate and complete project that ends with the issuance of a report on the feasibility of the idea.

## Return-on-Investment

Hiring a specialized company to do the feasibility study has many advantages, as mentioned. Also, consider the case of a company undertaking an investment project to generate money, such as developing a new product or building a hotel. In this case, it will be very important to assess the value of the investment by calculating what is known as the return-on-investment (ROI). ROI involves using mathematical formulas to calculate the amount of money spent on the project and the expected (forecasted) revenue over the years to assess wether the project will prove profitable and whether the anticipated revenue outweighs the expenses. If calculating ROI is needed and the company sponsoring the project has little knowledge of the calculation required for it, then it will be a good idea to hire a specialized company.

However, the return-on-investment is not always financial in nature, as not all projects are undertaken to generate profit. For example, when the project is going on a vacation, the return on your expenses will be the pleasure of having taken the trip. I have made a simple template for the non-financial return-on-investment which can be attached to the feasibility study template to give the idea more support, as you will see in the example below.

### Example: Constructing a Community Park

The following template represents the feasibility study (including the ROI) done for the construction of a community park. This template should be prepared by the person who came up with the idea.

I will start using the templates now. Note that they are numbered from 1 to 23. Filling out each template as you work through the example should serve as a description of its parts (that is, my intention is for the template to be self-explanatory).

Template 1

### Non-Financial Return-on-Investment

| Expected Return | Chance of Attaining the Return | Link with Organizational Goals | Cost |
|---|---|---|---|
| Provide new services to the community | High | High | Money needed for the project |
| Encourage people in the community to practice sports and help improve their health | Medium | High | Time dedicated to the project by the assigned project manager |
| Encourage tourism | High | High | Administrative activities to prepare the contract |
| Decrease crowding in other parks | Medium | High | |
| Preserve the environment through the planting of trees | High | High | |
| Increase the chances of the city's winning the Most Beautiful City in the World competition | | High | |

Template 2

*Feasibility Study*

| Description of the Idea |
|---|
| Construct a park that includes green landscapes and children's playgrounds in the northern part of the city |
| **Available Alternatives** |
| Not constructing the project, but it will lead to the loss of all the benefits described in the attached ROI template |
| **Availability of Funding** |
| An estimation of the required budget is around 4 million USD.  The finance department was contacted and funding can be secured if the idea is accepted by top management |
| **Can the Idea Be Implemented Technically?** |
| Constructing the park does not require any technological means that are not available in the market |
| **Availability of Human Resources to Run the Project** |
| Project manager will be provided by the parks department |
| **Conflicts with Other Projects** |
| No conflicts are present.  The roads department was contacted to make sure no roads are planned in the proposed park area |
| **Recommendations** |
| The idea was reviewed in the monthly top management meeting and approval was granted to start planning for the project.  The approval was granted and signed for by the: Finance Manager, Strategic Planning Manager, Manager of Roads and Manager of Parks departments on 16 April 2010. |

We assume that the templates were presented to a committee of the top management and were approved. Amanda can now start planning for the project as per the Project Book method according to which she will prepare a binder and fill it with all the templates specified in the Project Book framework.

**Benefits of Using These Templates**

1  You identified and documented the benefits expected from the project.

2  You started the process of turning the idea into a project.

3  You started thinking about the possible obstacles to the project.

4  You made sure that the project deliverables are in line with the organizational objectives.

# Project Charter (Project ID Card)

Once the feasibility study has been done and approval has been granted to proceed with the project, you can start the planning phase. You can consider the project charter to be the birth certificate of the project. It will contain the following essential information:

## Name of the Project

It is important to select for your project a meaningful name that will ease communication about it. Try to avoid long and complex names which might lead to the creation of different shorter versions of the original name by people dealing with the project, thus causing confusion. You might also give the project a reference number as per company procedures to link it to different systems, such as the finance and contract systems.

## Project Start Date

You can use the date project approval was granted as a result of the feasibility study as the project start date. You should, however, make it clear to people involved in the project that this date is for starting project planning, not execution.

## Selection of the Project Manager

Choosing the right project manager for the project will have a huge impact on whether the project will succeed or fail, as he/she will be the overall supervisor and decision maker. Selecting a project manager should be done through top management levels; doing so will give the project manager more power in managing the project and will ensure greater support when he/she must request help and resources from other departments within the organization. In many cases, the one who came up with the project idea will be assigned as project manager. In any case, the following should be taken into consideration when selecting a project manager:

• The individual's knowledge and experience in the field of project management;

• The individual's knowledge and experience in the technical field involved in the project;

• The individual's knowledge regarding the internal policies and procedures of the organization;

• The individual's ability to communicate and lead;

• The individual's availability, as he/she will have to see the project through from the initiation stage to closeout.

**Basic Goals of the Project**

The goals of the project are mentioned briefly in the feasibility study, but they might need to be amended after being reviewed in order to get the approval to proceed with the project. Now is your chance to document and "lock in" the original objectives of the project because, as the project progresses through its life cycle, sometimes the original ideas are forgotten and deviation from them occurs. It will be helpful to always be able to return to the project charter and check on whether the project is still aligned with its original goals.

**Approvals and Signatures**

The project charter is an official document and should be signed by top management. If possible, it should also be signed by the general manager. This piece of paper will be the access card for the project manager in asking for resources and justifying to others what he/she is doing.

**Example of Constructing the Community Park: Filling Out the Project Charter**

**Note:** The selected project manager should be the one to fill out this template.

Amanda did some brainstorming with the manager of the parks department and selected the name "REEM" for the park. The following is the project charter template that was prepared for the park:

Template 3

*Project Charter*

| Project Name | REEM Community Park |
|---|---|
| Project ID | 2-P-2010 |
| Expected Duration | 1.5 years |
| Expected Budget | $2 million US |
| Project Goals | Serve the people of the northern neighborhoods of the city by providing a park that includes green landscapes, walkways and children's playgrounds |
| Project Owner | The Parks Department |
| Project Manager | Engineer Amanda Smith |
| Start Date | 1-6-2010 |
| Signature | The GM |

**Benefits of Using This Template**

1  It officially launched the project.
2  A name has been selected and a project manager has been assigned to the project.

# Chapter Four
# Initial Planning

I have divided the planning stage into two chapters concerning the initial and final planning. I did this to separate risk management out as a chapter of its own, called "Final Planning." Risk management is so critical and important in project management because it questions the quality of all other planning elements and has the potential to change and amend them. When doing risk management, you will need to review all the planning that has been done. Once you finish risk management, you will be ready to start executing your project. A separate chapter is dedicated solely to it to emphasize its importance and also for organizational purposes.

The project planning stage may be the most important among all the project stages. Unfortunately, planning is sometimes ignored, as mentioned in the first chapter of this book. My intention in this chapter is to go through and explain the different planning elements and fill out the respective templates to produce what is called a "project plan." Having a project plan will help in that it:

- Will act as a road map we can follow during project execution. As I mentioned before, lack of knowledge surrounds projects in their startup stage; having a plan (or a map) containing all ideas, analyses and expectations will help the project manager navigate easily and confidently in executing the project.

- Will act as the benchmark against which to compare the actual project progress. This will enable us to determine if the project is on track or delayed, on budget or over budget and so forth.

**Note**: Some people might think the project management plan is just the diagram or chart that describes the tasks of the project over time. I would like to make it clear, though, that the project plan is much more than a mere time schedule; in face, a time schedule is actually just one part of it. Think of the project plan as a master plan consisting of many smaller plans, like a folder with many subsections in it. As per the Project Book framework, the project plan will consist of the following elements/sub-plans:

- The Project Team Plan;

- Stakeholders Management Plan;

- Information Management Plan;

- Project Scope;

- Project Map;

- Project Quality Plan;

- Project Time Plan;

- Project Budget and Resources Plan;

- Contractors Management Plan;

- Change Management and Progress Reporting Plans; and

- Risk Management Plan.

In the following sections, we will cover each of the elements/sub-plans.

# The Project Team Plan

Chances are you will need people to help you (the project manager) in running the project. Those people will make up the project team. In an organizational context, you will be lucky if you have direct authority over them and caan assign them tasks and monitor and assess their performance. But in some cases, they might not be under your direct authority and you will need to expend some effort trying to get them to cooperate with you. Let's discuss the different situations and what the relationships might be between the project team and project manager:

**First Situation:** The project team is under the direct authority of the project manager.

This is the best situation, where members of the project team are expected to give full attention and commitment to the project. This situation is possible when the project manager holds a management position in the company and uses staff who are already under his/her supervision. Another situation is when the organization's top management officially assigns a team for the project whose members are instructed to give full attention to the project and follow the project manager's initiatives.

**Second Situation:** The project team is not under the direct supervision of the project manager.

This situation is mostly present in organizations that follow a traditional structure of distributing staff over separate departments (called functional organizational structure). In this case, the members of the project team will be collected from different departments and will be required to work both under their functional manager and under the project manager; in most cases they will tend to focus more on their normal work than on the project activities (because their yearly evaluation is based on how well they do their work). Here, the project manager should try using different ways of securing commitment from the project team. And since the project manager is not considered the formal authority over the project team, one of the best ways of winning their interest is to help them see the project and its goals and objectives as beneficial to them and the company. Also, the project manager must establish communication with the functional managers of the project team members and reach an agreement with them that they will direct their staff to devote sufficient time to the project.

**Third Situation:** There is no project team (that is, the project manager will manage the project by him/herself).

This situation occurs when the project manager works in an organization that uses a consultant and a contractor to design and construct their projects. Of course, your project (and you) will benefit from a project team even if it is only temporary. For that, I advise you (in case you are in this type of situation) to agree with a few of your co-workers to "act" as team members in each other's projects. By this I mean that you will cooperate with your colleagues who will help you in filing out the templates and brainstorming for your projects on a voluntary basis, and you will do the same for their projects.

## What Do We Expect of Each Other?

Being the project manager (with the greatest and most direct accountability for the project's success or failure), you will expect your project team members to give the project their time and effort in all possible ways to ensure its success. Also, you will expect them to work as a team in solving problems and avoiding conflicts (especially of a personal nature), and to share information among the team. On the other hand, your project team members will expect you to be fair in assigning tasks to them and assessing their performance, and to share relevant information about the project with them.

## How to Improve the Productivity of Your Project Team

- **Encouragement**

Some of the strongest encouragement can be given through "non-financial" rewards, like thanking a team member for a good idea in front of other members. In the course of my work, I have noticed how the simple use of verbal acknowledgment can make people happy and want to give all they have for a certain cause. So make it a habit to always commend good work where deserved.

- **Training**

Try identifying any special requirements of the project, and then match them (via training) with some of the team members. Such specific training will improve the progress of the project, will decrease frustration associated with lack of knowledge and will, in many cases, improve the morale of the team members as they acquire new skills and knowledge as a result of the project. An example of such training is teaching a team member to use a specialized software or training a few memberson how to conduct quality audits.

- **Sharing Information**

Updating the project team with new information will improve trust between them and the project manager and will give them a sense of belonging to the project (as there will be no hidden information). Also, knowledge of the latest project updates (such as change requests) will prevent the team members from being put in embarrassing situations that may result from misinformation or lack of knowledge on project progress.

## Leadership Styles

I will not go into detail about leadership style, as it is a deep and complex subject. However, I want to describe two situations that are frequently repeated in projects and demand different forms of leadership:

**First Situation:** When the majority of the project team is composed of fresh graduates. In this case, it is better for the project manager to follow an approach of instructing and coaching team members on what they should do since they lack experience. The project manager should closely monitor their performance and decisions to decrease the probability of mistakes.

**Second Situation:** When the majority of the project team is composed of experienced staff. In this case, the best thing for the project manager to do is to give his/her team more space to do their work with minimum intervention. Decisions should be discussed with the team in a more interactive way than in the previous case, since it is highly likely that team members will have comments and would feel unappreciated if left out of the decision-making process.

## Selecting the Project Team

If you can select your own project team, consider the following:

- The individuals' nowledge of and experience in project management and/or the technical aspect of the project;

- Availability of the person/s;

- Ability of the individuals to work in a team setting.

## The Project Team Template

The template we are going to use is very simple. It is concerned with recording the names of the team members and their assigned tasks. However, simple as it might look, it will save you from trouble arising from conflicts between team members, and you will not have to be surprised by words like "I don't know what my role is."

**REEM Park Example: Project Team Plan**

Due to the lack of direct staff under the supervision of Engineer Amanda, she arranged for some of her co-workers to help her on the project, especially in brainstorming and planning. She also asked the department manager to assign a secretary to help her in the administrative work of the project. The following is the filled-out template she made:

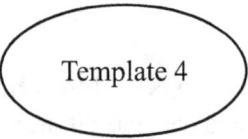

Template 4

*Project Team*

| No. | Name | Tasks | Signature |
|-----|------|-------|-----------|
| 1 | Amanda | Project Manager: Responsible for completing the project book binder and all aspects of the project during the different stages. Directly responsible for managing the consulting and contracting company. | *amanda* |
| 2 | Ali | Help in completing the project book binder | *ali* |
| 3 | Sara | Help in completing the project book binder | *sara* |
| 4 | John | Help in administrative work like preparing documents and memos and calling for meetings | *john* |

- Starting the project planning stage by selecting the project team is essential as you need people to help you brainstorm as you fill out the templates. Remember that a key element of the Project Book model is the involvement of people who invoke discussion of different ideas and, thus, reduce errors in judgment by preventing the occurrence of a single point of view.

- Note that team members are to sign the template beside their assigned tasks to show that they understand and accept them.

- If you discover more tasks along the way, go back to this template and incorporate them into the task list of the appropriate team member/s.

**Benefit of Using This Template**

You were able to select the team members for your project and clearly assign tasks to each one of them.

# Reviewing Previous Lessons Learned

No matter how big or small it was, any past project should have added something to the collective knowledge of an organization or an individual. The difference between lessons learned and information you might find abundant in books and articles is that the former are specific to the organization while the latter are general. Information that is specific to the organization will identify specific areas of improvement that should yield faster and more direct results, unlike general information, which, in some cases, won't even apply to a certain organization or setting. Our goals for reviewing past projects are:

1  To review comments and suggestions raised from past projects and the possibility of using them in the project at hand;

2  To identify mistakes that occurred and obstacles that arose in the past and try to avoid them;

3  To identify opportunities that were wasted and how they could be harnessed now;

4  To complete the cycle of knowledge transfer by actually using experience and wisdom learned from the past.

**Where to Find Lessons Learned**

You may look for lessons learned from projects in

- The project management office;
- The knowledge management office;
- The department conducting the project;
- Organizational archives.

Finding lessons learned might be a grueling exercise, especially in companies which are just starting the application of project management principles. In many instances, you will not readily find lessons learned organized in reports (sometimes the term "lessons learned" will be alien to an organization, let alone implemented). However, you shouldn't give up. You can look for lessons from past projects in a variety of project files and reports, especially progress and risk reports.

In case you are not able to find written information about past projects, you can arrange for a meeting and invite project managers of previous projects. This will be called a "memory" session to document comments or important aspects they remember about the projects they handled in the past.

**REEM Park Example: Reviewing Lessons Learned**

Amanda did not find any readily available reports about previous park projects done in the department. Also, since the department had moved to a new location during the past month, all its project files had been sent to storage in the industrial area of the city and Amanda thought it would be very difficult to locate them. Amanda decided to call for a meeting with project managers who had managed similar projects in the past, and some lessons learned were recalled. The following template shows past lessons learned and how they can be incorporated in the current project:

Template 5

*Lessons Learned Review*

| No. | Previous Project Name | Lesson Learned | How to Use It |
|-----|----------------------|----------------|---------------|
| 1 | Central Park Project | The consulting company made many mistakes in the design drawings which required rework and delayed the project. | Exclude the consulting company from participating in REEM project |
| 2 | Central Park Project | The children's equipment used was missing many parts when it arrived from the manufacturer. | If equipment from the same supplier will be used, instruct the contracting company to double-check and make sure that all parts are shipped together. |
| 3 | Lake Park Project | A new irrigation system was used which proved both cheap and reliable. | Advise the consulting and contracting company to look into using the same irrigation system. |

**Benefits of Using This Template**

1  You reviewed past mistakes so you could reduce the chance of their recurrence.

2  Using the template (that is, its availability) assures the company that the project manager reviewed the organizational knowledge in the lessons learned and, thus, completed the life cycle of knowledge management.

# Stakeholders Management

If we divide project management into the old and modern types, then the use of stakeholders management principles will distinguish modern project management. Stakeholders management urges the project team to consider internal and external forces acting on the project and helps them realize they are not in absolute control and must consider the needs of others.

**What Do We Mean by "Stakeholders"?**

Stakeholders are the people (organizations) who are interested in your project. Let us analyze the word "interest." Someone might be interested in your project for a variety of reasons; for example:

- Your project is within the administrative jurisdiction of a governmental agency (interest here involves the law);

- Your project has a positive impact on a group of people, such as when you construct a park in their neighborhood (interest here stems from the desire to benefit from the deliverables of the project);

- Your project provides work opportunities for other companies, such as consulting firms (interest here comes from the desire to make money);

- Your project has a negative impact on a group of people, such as in the case of the demolition of a building in the middle of the city, resulting in noise and lack of parking space for residents of neighboring buildings (interest here is to avoid or decrease the negative impact perceived to be caused by the project).

The above are only some examples of reasons someone might be interested in your project. Note that interest might be direct or indirect and it might be positive or negative. However, be assured that if someone is watching your project closely, he/she/it will try to influence it and pull it toward achieving his/her/their interest. Your role as a project manager is to acknowledge that such interests exist and try to steer them for the good of the project.

**Examples of Stakeholders:**

- The project sponsor (the most important since he/she provides the funding and the project is done to achieve his/her goals);

- Governmental entities;

- Local people;

- Private companies;

- Media;

- Departments and sections in the organization running the project;

- Top management of the company running the project;

- Neighboring countries/states/provinces/municipalities.

**Why Should Project Managers Care about Stakeholders?**

As a project manager, you want to decrease external and internal influences on your project since they may hinder its success. However, no matter how hard you try, it is impossible to shield your project entirely from such influences. As such, it will be easier for you to be proactive and identify possible influences beforehand and try to control their effects on the project through a set of possible actions, such as trying to incorporate the influences, steering them away from the project or striking a balance between them and the well-being of the project. In short, the project manager should try to reduce the negative effect of stakeholders' influences by trying to satisfy/address their requirements early on.

**Examples of How Stakeholders Can Affect Your Project**

**Example 1:** A project to construct a football field was delayed because the people living close to the proposed location complained to the municipality that it would cause noise and crowding in the area.

**Example 2:** A project to create a website for a department in a company was delayed because the IT department wasn't involved in selecting the designing company.

**Example 3:** The budget for a project to construct a public water fountain was increased due to a change in building materials as per the request of a senior manager who wasn't consulted in selecting the materials.

**Example 4:** A project faced difficulties in securing some needed materials because the suppliers did not have enough on hand (they had not been informed beforehand of the volume needed for the project).

**Example 5:** A project to construct a water dam in a river was canceled because a neighboring region complained that the dam would decrease its share of the water supply.

**Example 6:** In a project to maintain a school building, a rich man who had attended the school in the past donated money to refurbish and equip the school library.

**Note:** From the examples above you will notice that the different motives of stakeholders might not always be logical, but what might not be logical to the project team might seem perfectly reasonable in the mind of a stakeholder. In stakeholders management, it is not our goal to assess the logic of stakeholders' requests, but to incorporate them in the project in a manner that will decrease any negative effects on the project.

**What Is Stakeholders Management?**

As a project manager, you will conduct the following activities under the umbrella of stakeholders management:

1 Identify the stakeholders in your project.

2 Prioritize stakeholders as per their interest in and ability to influence your project.

3 Identify the potential needs of the stakeholders.

4 Put a plan in place to satisfy stakeholders' requirements as per project scope and resources.

These activities will be discussed in detail in the following section.

**Stakeholders Management Template**

This template might seem a little different from other templates used in this book, so I will describe it in detail here. It consists of two items, the first of which will help you in identifying stakeholders and prioritizing them. The second will help you in thinking about their needs and the best ways of achieving them. The following is a description of the sections that make up the template:

**1 Identifying Stakeholders**

To identify stakeholders, invite your project team to a brainstorming session and use the following approach:

Use a big board (or a piece of paper) and write the following guide words:

- Types of stakeholders (individuals and organizations);
- Location of stakeholders (inside the company or outside);
- Type of interest (positive or negative).

Then try to give examples for all the above types.

## 2 Stakeholders Prioritization

Prioritization is a recurring theme in project management and in management in general. When you prioritize, you identify elements that are more important than others; hence, you can utilize your time and resources to respond to needs based on the hierarchy of importance. When prioritizing stakeholders, a project manager should look into two conditions: the ability of a stakeholder to make changes in the project and the degree of interest of that stakeholder in using his/her strength.

A famous method is to arrange stakeholders into four types (which translates into quarters on a piece of paper) based on their importance. The first section will list the most important ones, then the second most important, the third and, finally, the fourth. The four quarters that prioritize stakeholders compose a stakeholders analysis matrix which looks like this:

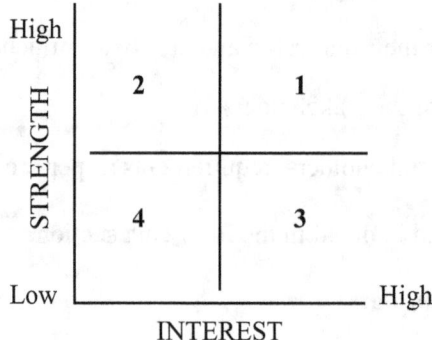

**First Quarter:** The strength of the stakeholder is high, and the interest in using the strength is high. **Second Quarter:** The strength of the stakeholder is high, and the interest in using the strength is low. **Third Quarter:** The strength of the stakeholder is low, and the interest in using the strength is high. **Fourth Quarter:** The strength of the stakeholder is low, and the interest in using the strength is low.

It is difficult to be objective in prioritizing stakeholders because deciding on the level of strength and interest of a stakeholder will vary from one person to another. To improve your output in this exercise, you should do it in a team setting and consider the varying opinions of different people.

Once you have prioritized your project stakeholders, you will have a clearer picture of how to distribute resources to accommodate their requests. For example, if you receive two requests at the same time from two different stakeholders where one of them lies in the first section of the prioritization matrix and the second lies in the third section, you can comfortably start by responding to the request from the first stakeholder.

## 3 Responding to Stakeholders' Requests

Now you want to think of potential requests/requirements that will be made by stakeholders. In a perfect scenario, you would want to respond to all requests from stakeholders. However, it will prove very difficult because:

- Some stakeholders' requests might interfere with the project scope;
- Some requests might be too costly to satisfy;
- Conflicts may arise from different requests made by different stakeholders.

Before responding to any request, keep in mind that not satisfying a request is an available option you can consider. As a project manager, you should always weigh and balance the costs of responding to a request with not responding, and the potential effect of your decision on the project.

### REEM Park Example: Stakeholders Management

Amanda filled out the stakeholders template using the following procedure:

**First:** The project team brainstormed concerning the list of potential stakeholders and came up with the following:

Template 6

*Stakeholders Identification and Prioritization*

| Stakeholders | | |
|---|---|---|
| GM | | |
| Parks Department | | |
| Roads Department | | |
| Environmental Agency | | |
| Local People | | |
| Private Companies | | |
| Potential Investors | | |
| Media | | |
| Tourism Authority | | |

STRENGTH — High / Low

INTEREST — Low / High

2    1

4    3

**Second:** To prioritize the stakeholders, the project team went through the list of identified stakeholders and, for each one, asked the following questions:

1  What is the strength of the stakeholder?

2  What is his/her/their interest in influencing the project?

The template was then finalized as follows:

Template 6

*Stakeholders Identification and Prioritization*

| Stakeholders | | |
|---|---|---|
| **Stakeholders**<br>GM<br>Parks department<br>Roads Department<br>Environmental Agency<br>Local People<br>Private Companies<br>Potential Investors<br>Media<br>Tourism Authority | High<br><br>Media<br><br>**STRENGTH**<br><br><br>Low | GM<br>Parks department<br>Roads Department<br>Environmental Agency<br>Tourism Authority<br><br>Local People<br>Private Companies<br>Potential Investors<br><br>High<br>**INTEREST** |

**Third:** After prioritizing the stakeholders, the project team started to think of their potential requirements and how they might be satisfied, as illustrated in the template below.

**Note:** Here it will prove beneficial that the previous template helped us prioritize the stakeholders by urging us early on to start thinking about the potential requests of important stakeholders.

Template 7

### *Stakeholders Requirements*

| Stakeholders | Potential Requirements | How to Satisfy Requirements |
|---|---|---|
| GM (General Manager) | Project is completed within time and budget. The design should be top notch | Through proper planning and selecting a good consulting firm |
| Roads Department | Submit the design sketch and drawing providing for parking spaces and connecting roads | Through proper communication by project manager |
| Environmental Agency | Provide information and 3D drawings about the project progress | Through proper communication by project manager |
| Local People | Provide information on the project progress. Provide sufficient parking spaces. Design the roads in a way that will not increase traffic jams. | Provide information through the public relations department |
| Private Companies | Provide a clear description of the requirements of the park once the request for tendering is advertised in the newspaper | Proper planning |
| Potential Investors | Provide buildings with adequate spaces to be allocated for shops and restaurants | Include this requirement in the tendering document for the consulting firm designing the park |
| Media | Provide information about project progress | Through proper communication by project manager |
| Tourism Authority | Provide information and 3D drawings about the project progress | Through proper communication by project manager |

**Note 1:** This template should be consulted and reviewed at different stages of the planning process to ensure that the requirements of stakeholders are incorporated into the project as far as reasonably practical (for example, when writing the project scope and determining quality requirements).

**Note 2:** As you saw from the template, one of the best things to do for your project's stakeholders is to provide them with the information they require; this will usually give them a sense of importance.

**Benefits of Using These Templates:**

1 You identified potential people/organizations interested in your project.

2 You prioritized stakeholders so you can respond to their requests in an organized manner.

3 You identified potential requirements of the stakeholders so you can try to satisfy them and incorporate them during project planning.

# Information Sharing Plan

Getting the right information at the right time will increase your knowledge and help you succeed in what you are doing. On the other hand, lack of information (and I don't mean its non-existence, but rather, not sharing it) will lead to confusion and conflict. In the table below, you can see some examples of the negative effects of not sharing information:

| Event | Impact |
|-------|--------|
| Not sending the monthly report to an important stakeholder | The stakeholder becomes angry and tries to intervene in the project to show his importance |
| A change request is sent through a fax machine that is not used frequently by the contracting company | The change request is not discovered in time, which leads to its not being implemented |
| Holding a meeting to discuss the contracting document but forgetting to invite a representative from the legal department | The quality of the discussion will be affected as the legal department might have had valuable input to offer |

**How to Make an Information Sharing Plan**

You can make an excellent plan by answering the following questions:

**1 What is the information that must be shared?**

Note that the question is not what kind of information you want to share. We can rewrite the question as: What is the information that, if not shared, will cause trouble to your project? Here are some suggestions:

• The project scope;

• The risk register;

• Project progress report;

- Change requests;

- Product design.

## 2  Who will receive the information?

You will have to determine the people who are interested in information about your project, and those who are not interested in the information themselves but will want to receive it anyway as an indication of their importance. You can select these people from the list of stakeholders and try to match them to the list of information you identified by answering the first question above.

## 3  How do you share the information?

Nowadays we have different means of sending information. The right way will be determined by different factors, such as the available technology (for both the sender and receiver), the geographic area the project spans and the speed required for delivery. The following are examples of the different ways of sending information:

- Using normal mail;

- Using a fax machine;

- Using e-mail;

- Uploading information onto a website;

- Using the phone.

To select the best ways of sending information, note that information can be official or non-official and written or verbal. A general rule is that important information should be sent officially in written format. For example, change requests for the project must be written in an official letter and tracked to ensure delivery.

## 4  What are the required meetings?

Meetings are part of information sharing, and we need to make sure that all required people are invited to them. So the project manager should try to identify meetings for his/her project beforehand, prepare a list of people for every meeting and send invitations beforehand. The list of all meetings with the required people can be prepared and sent to all concerned parties to be posted in their calendars.

## 5  When should you send information?

Information must be sent early enough to give the one who receives it the time to analyze it and respond if necessary. In the information sharing plan, it is advisable to set a date for sending information because it will create a sense of commitment about sending it on the part of the project team. Here are some examples of how to set dates for sending information:

- Information is sent on the fifth day of every month;

- Information is sent two days after each meeting;

- Information is sent on the same day it is collected.

## 6  Who will send the information?

The project team is responsible for sending information related to the the project. The project manager should oversee this task and assign (when possible) someone from the team to execute it.

### Skills That Help in Communication

Sharing information demands communication with others. These are some considerations while communicating:

In verbal communication (like phone calls, presentations and meetings),

- Always start with an introduction about the agenda of the meeting;

- Watch your tone of voice. Your voice should be loud enough to be heard, but it shouldn't carry meanings you don't want to send out since sometimes the tone of voice can imply a negative meaning (for example, attack, blame);

- Watch your body language. You might be able to control your spoken words, but as shown in many studies, body language conveys more meaning when it comes to the messages you are sending;

- Don't forget that listening is as important as speaking, so give others time to ask questions or offer ideas;

- Respect customs of people you deal with which might be different from yours (for example, customs in greeting one another);

- Listen to understand the viewpoints of others, not to find mistakes.

In written communication (like reports and email),

- Don't forget to put a title/subject line into your message;

- Register your correspondences for future tracking;

- Use simple language and stay away from abbreviations as much as possible;

- Make sure the person receiving your message can open it. For example, if you are sending an email attachment, make sure the recipient has the software to open it;

- Check your writing for spelling and grammar;

- Track your correspondence to make sure it arrived at its intended destination.

**REEM Park Example: Information Sharing Plan**

Engineer Amanda and her project team asked themselves the questions presented in this section and came up with the following template:

Template 8

*Information Sharing Plan*

| Information | Receivers | Method of Sending | Frequency of Sending | Who Will Send the Information? |
|---|---|---|---|---|
| Risk Identification Meeting | Project team + representative from the roads and contracts departments + other project managers from the parks department | Email | One time | John |
| Time plan | Project team + project management office + consultant and contractor | Official letter | One time and after any change/s | John |
| Kick-Off Meeting with the consultant | Project team + manager of the parks department + project management office | Email | One time | John |
| Preliminary design of the park | Project team + the manager of the parks department | Email | One time | John |
| Final design of the park | GM + project team + environmental agency + tourism agency + media | Official letter | One time | Amanda |
| Project progress report | Project team + finance department + PMO + manager of the parks department + contractor + consultant + the manager of roads and contracts manager | Email | First week of every month | Amanda |
| Lessons Learned Meeting | All available stakeholders | Email | One time | John |

| Receiver | Contact | | Receiver | Contact |
|---|---|---|---|---|
| GM | Secretary | | Environmental Agency | Through public relations |
| PMO | Email | | Roads Department | Email |
| Tourism Authority | Through public relations | | | |

**Note 1:** You can use this template as a tracking tool by adding a column for tick marks to be made every time information is sent.

**Note 2:** Before filling out this template, you should review the stakeholders' requirements to identify any need to send information to a particular stakeholder. By doing so, you will make your templates more realistic and more practical, which is actually the intention of the Project Book methodology.

**Benefits of Using This Template**

1  You thought about and identified the information you need to share.

2  You identified the people who will receive the information and how you will send the information to them.

3  You reviewed the stakeholders template to identify any requirements to send information to a particular stakeholder.

# Project Scope

"Project scope" is a widely used term in project management that means the description of the project deliverables. This description of the project product or output will help us in identifying the tasks we need to include in the project in order to achieve it.

All the templates we have filled in until now will support project planning, but we have not yet started looking into the product of the project and how we can actually achieve it. Having a clear idea of the project doesn't mean that it will be easy to write down the project scope. Take the example of REEM Park: The idea is very clear and simple and can be simply summarized as constructing a public park to serve the people of a certain part of the city. However, you may notice that the idea can be explained in many different ways, such as:

• The park will/will not have children's playgrounds;

• The park will/will not contain restaurants;

- The park will contain a big fountain near the entrance;

- There will be an artificial river in the park;

- The park will have a fish aquarium.

Observe that we can not confirm or exclude any of the descriptions given above by using only the project idea. Having different possible descriptions for a single project is a potential risk because we are left unable to identify the required tasks; as a result, we cannot plan for the project. For example, constructing a park with an artificial river is different from constructing a park without one. As you can see, it is very important to transfer the project idea into a clear description of the desired project outcome. This is our task in the project scope.

**How Does One Define the Project Scope?**

The project team must not rush into designing the project scope because it will be the basis of all subsequent planning elements, and any change in the scope will lead to a lot of reworking of the plan, which can lead to mistakes. To create the project scope, do the following:

**First:** Have a clear idea of the reason/s for starting the project. To know this, review the templates for the feasibility study and project charter.

**Second:** Study the requirements of the project sponsor.

During the stakeholders management exercise, we tried to think of their potential requirements so that we can try to satisfy them. However, the project sponsor is the most important stakeholder, and his/her needs should be confirmed in a more direct way through meetings and interviews. In the case of projects run for organizations, the project team must meet with senior management to identify what they want the project to deliver. Collecting the sponsor's requirements is not always easy as he/she might not know exactly what he/she wants, especially if the project involves certain technicalities that he/she is not familiar with.

Consider an example in which the human resources manager wants to install an electronic archiving system in his department. In this case, the HR manager might have a rough idea of the project but will likely not be able to specify the number and type of scanners needed. Here the project manager must sit with and interview the HR manager to determine his actual requirements and may ask him/her questions such as:

- How many documents are produced daily?

- How many documents are in existence in the HR department?

- How many people will use the system?

- Do the documents contain drawings?

- Is there a need to search inside the documents once scanned?

Consider the interview with the sponsor as a mini data collection exercise which might result in a description of the project that is different from what is in the sponsor's mind. If indeed the interview produces a different description from what the sponsor had in mind, then the project manager should build on the new description and consider this a success for the project, as the project will be able to avoid trouble resulting from conflicting views between the sponsor and project team.

**Third:** Identify different views of the project.

I advise the project team to conduct a quick survey among some of the stakeholders and ask them what they think of the final product of the project. The team will benefit by identifying features of the final product that are not in line with the sponsors' requirements and, as such, can be explicit in rejecting them. For example, for REEM Park, some stakeholders might mention that the park will contain bird cages which are not specified by the sponsor, so in writing the project scope, it will be mentioned clearly that the park will not contain bird cages. This is an important step, as many troubles facing projects are generated by assuming that the project deliverable will contain something that isn't there.

**Fourth:** Use the project scope template.

This template is made up of two useful exercises that will help in completing the project scope as follows:

- **Project Picture**

You don't need to be a great painter, but if you draw the final product of your project, it will help you illustrate what words fail to and should stimulate your thinking to uncover hidden components of the final product.

- **What's In and What's Out?**

Here you will actively list what is not included in the project product. It is a very helpful exercise since, usually, people only think of what is supposed to be included. Thinking of the opposite should provoke their thinking in new ways.

**Fifth:** Secure the approval of the project sponsor on the scope.

It is worth mentioning that it will be a waste of time to continue planning for the project if the project sponsor doesn't accept the project scope (by signing the scope template). So, if the project manager doesn't present a clear scope that is agreed upon by the sponsor, then chances are there will be many changes to the scope down the road.

As you have perhaps correctly figured out, the ultimate goal in defining the project scope is to "lock" it in so as to decrease the possibilities of its being changed. It is very difficult to prevent changes to the project entirely (and sometimes changes are beneficial), but by having a signed scope template, one should be able to decrease the number of changes because a considerable amount of time has been spent in thinking about the best way to describe the project scope.

**REEM Park Example: The Project Scope**

Following the steps discussed in this section, Amanda and her team filled out the project scope template as follows:

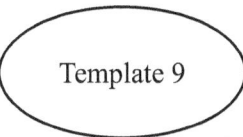

Template 9

*Project Scope*

**Project Picture**

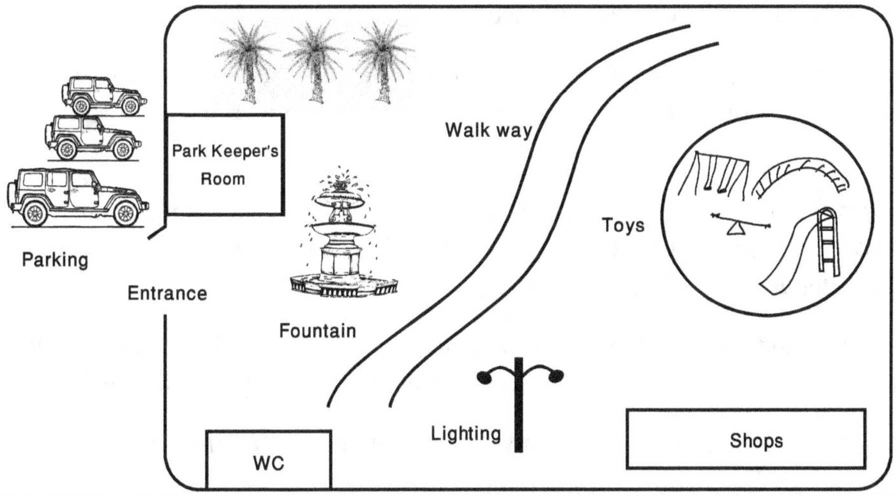

Parking

Park Keeper's Room

Entrance

Fountain

Walk way

Toys

Lighting

WC

Shops

**What's Out**
- Parking spaces
- Connecting roads
- External lighting
- Football and volleyball fields
- Swimming pool
- Bird cages

**What's In**
- Green landscapes
- Internal facilities, WC, shops, theater
- Fence
- Internal roads
- Running pathways
- Children's Toys
- Internal lighting
- Fountain

**Benefits of Using This Template:**

1  It provoked you into thinking more about the deliverables of your project.

2  You identified what's included and what's excluded in your project.

3  You reviewed some previous templates.

4  You created a summary template of what your project output will look like.

# Project Map (Breaking Down Your Project)

One of the most important things in planning your project is to break it down into its basic components because this will enable you to identify the required activities. Once you know them, you can assign money, time and quality to them.

### The Goal of Breaking Down Your Project

Your goal is to construct a map for your project showing all the tasks (and only the tasks) that you need to undertake to achieve the final output. Having a map of your project will enable you to:

• See the whole project on one piece of paper, which will ease the process of communicating what your project is about to others. Also, by seeing all your project tasks together, you will have a better understanding of their complexity and interdependency;

• Assign time and resources to your project more easily and with greater accuracy. This is because dealing with parts of the project is easier than dealing with the project as a whole;

• Identify missing tasks more easily and conduct better risk identification.

### How to Break Down Your Project

1  First, remember that the goal of breaking down a project is to identify the tasks so it will be easy to assign time and resources to them.

2  Start by making a square and writing the final product of the project in it (let's use the simple project of preparing a dinner for your friends):

**Dinner**

3 Think of the most essential components of the final product and write them in squares that connect to the final product as follows:

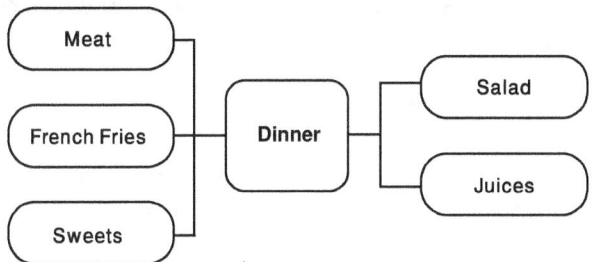

4 Now break down the major components into the tasks that will be needed to accomplish the desired result. It might help to ask what should be done to produce these components:

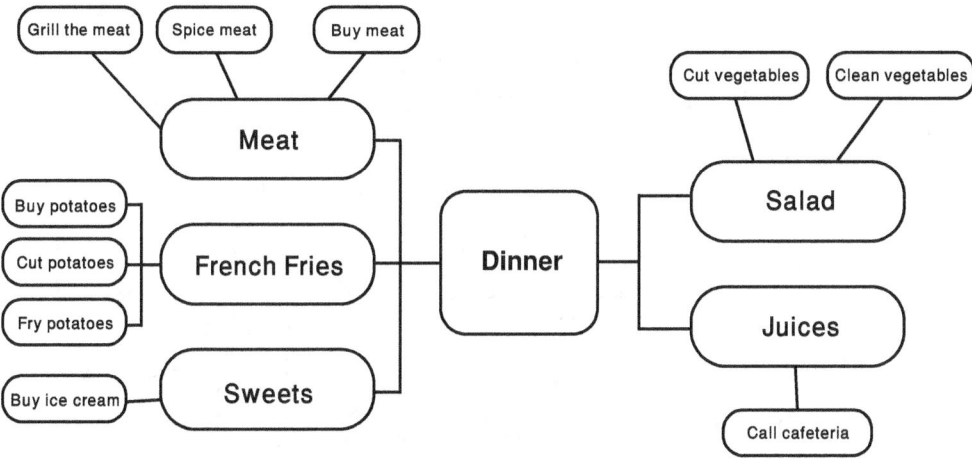

**Comments:**

1 By this time, you should be able to see the benefits of the project map. You can immediately identify the tasks that require money (like buying meat) and the tasks that will utilize the help of others (like preparing the juice) and the tasks that will involve special resources (like oil to fry the potatoes).

2 By considering all the squares in the project map, you will reduce the chance of mistakes that may result from forgetting some tasks.

3 Arranging tasks in a sequence is not important in developing the project map, as you will arrange them when making the project time plan.

4 You can stop breaking down a component when you believe that you have arrived at a task to which you can assign resources and time accurately and confidently. For example, you don't need to break down the task of cutting the potatoes into subtasks like bringing the knife, the cutting board, etc.

**REEM Park Example: The Project Map**

After finalizing the project scope, Amanda and the project team started to brainstorm on a whiteboard regarding what tasks should be included in the park project. After that, the project map was placed onto the template as follows:

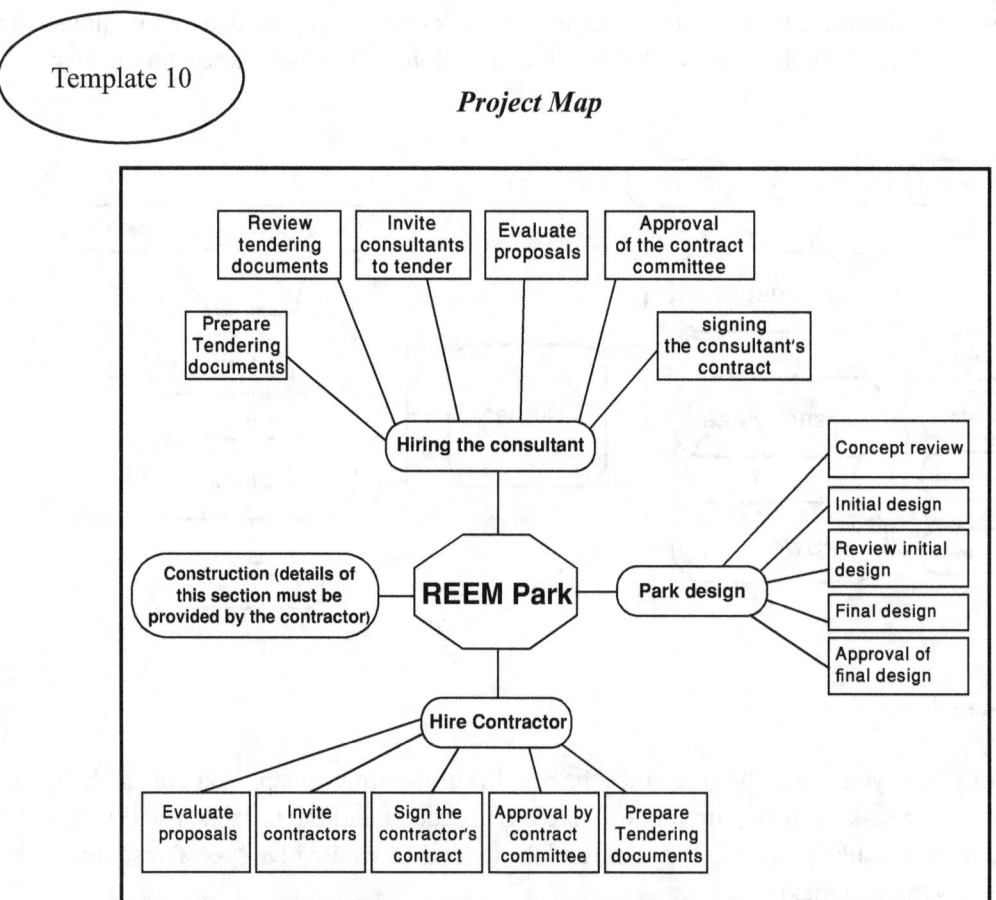

Template 10

*Project Map*

Note that the REEM Project was divided into four major components, as is the procedure for such projects in the municipality:

1 Hiring the consultant;

2 Designing the park;

3  Hiring the contractor;

4  Construction.

The components of the park project might be different from one company to another and from one location to another, but these components are frequently present in large projects (especially those run through government entities) where a consultant (to design) and a contractor (to construct) are involved.

**Benefits of Using This Template**

1  You created a project map that will enable you to proceed with ease in project planning, as will be shown in the following sections.

2  While making the project map, you reviewed the project scope and made sure your map covers all tasks required for the project (notice that, with each template, you are becoming more confident in your planning activities as past templates get reviewed and built upon).

# Quality Plan

Time, money and quality are among the most important elements in any project. In the Project Book, we will start by identifying the quality requirements before we make the time and budget plans. The reason we are starting with quality is because both the duration and cost of tasks depend heavily on the desired quality requirements.

Recall that a project is a means of producing a product, but similar products might have different characteristics and typologies. Take the example of having hundreds of different car models with different options, all sharing the same generic product name (that is, car). So, what would make you choose to buy one car instead of another? The choice you make is based on a collection of options and characteristics you think best suit your needs, or what might be called your desired quality requirements.

Usually, the word "quality" is synonymous with expensive, but it is not always the case, for quality in project management is concerned with how well a product satisfies a certain criteria. As a result, a project manager must identify all quality requirements which usually fall under the following categories:

## 1 Requirements of the Project Sponsor

The project is conducted in the first place to satisfy the needs of the project sponsor because he/she provides the project funding. Similar to identifying the sponsor's requirements while accomplishing the project scope, a project manager needs to sit down with the sponsor to collect the requirements (features) expected of the project product. To differentiate between the two, the requirements that were collected for the project scope are more of a description of the product, while the quality requirements are more about how the product would perform what it is intended to.

## 2 Governmental and/or International Requirements

The project manager must identify the technical requirements and standards the project must satisfy, which vary from one field to another. For example, when building a shopping mall, it might be a governmental requirement to provide specialized parking areas and access for people using wheelchairs. Another example is the requirement for firefighting facilities set forth by the civil defense authorities.

A good project manger will also try to anticipate any future rules and regulations that might affect the project and provide some allowance to accommodate them. For example, in adding an x-ray section to a hospital, the project manager discovered that the hospital has the intention of joining the international commission for hospitals (JCIA) and, as a result, had to do some research on their requirements for buildings. He discovered a need to provide separate changing rooms for patients under the privacy element of the JCIA standards, so he had to provide for them in the new section.

Identifying quality requirements is not easy because it takes a long time and may require the review of many documents and regulations. However, we shouldn't bypass it as there is little use for a product of poor quality (how about a website that crashes every day or a residential building that doesn't meet safety requirements?).

I mentioned that quality isn't necessarily determined by higher cost, but sometimes it might be so. Consider a project to build a luxury hotel. Here you need to buy expensive furniture for the hotel since it will surely be one consideration for measuring quality.

### Determining and Measuring Quality Criteria

To manage quality in our projects, we need to identify quality criteria (requirements) and know how to measure them. The following table is an example of some quality criteria in projects:

| Project | Example of Quality Criteria |
|---|---|
| Construct a customer service building for a service company | Size of the reception room<br>Number of parking spaces |
| Building an airplane engine | Weight of the engine<br>Fuel consumption<br>Working hours between scheduled maintenance |
| Creating a database for workers in a company | Number of workers that can be added<br>Possibility of accessing the system through the internet |

Note that once the quality criteria are determined, we can add some form of measurement to help determine whether quality is met or not.

**How to Find Quality Criteria in Your Project**

1 Use the project map and review it for any changes or updates.

2 For each task, determine the quality requirements of the sponsor.

3 For each task, determine any governmental or international quality requirements.

4 Determine the measuring criteria for the quality requirements and include them in the quality template.

**Exercise**

Write down the quality criteria and measurements for the dinner project presented in the previous section.

Using the steps mentioned earlier and referring to the project map of the dinner project, we can fill in the following table:

| No. | Task | Quality | Measure |
|-----|------|---------|---------|
| 1 | Buy meat | Use fresh meat | Date |
| 2 | Spice meat | Leave the meat marinated enough to take over the taste | Duration of marination |
| 3 | Cook meat | Cook until well done | Until red from inside |
| 4 | Buy potatoes | Suitable size to make french fries | Size medium to large |
| 5 | Cut potatoes | Cut in equal length | Same length |
| 6 | Fry potatoes | Fry until gold and crispy | From 7-10 minutes |
| 7 | Buy ice cream | Good brand | Specify the brand name |
| 8 | Wash vegetables | Remove all dirt | Cleanness of the vegetables |
| 9 | Cut vegetables | Size of the vegetables | Small-medium |
| 10 | Calling the cafeteria to prepare the juices | Give clear instruction to expedite delivery and avoid mistakes | Clear voice and fullness of information |

Determining the quality requirements for the project tasks and components will help us make better decisions regarding the time and resource requirements. In the dinner example, we decided that we need the meat to be cooked until well done, so we should allow sufficient time for that in the time plan. Also, we decided the quality of the ice cream should be high, so we should provide enough money to buy the ice cream.

> Quality management is very important for your project, and there are many training courses offered about it. It would be a good idea to nominate one of your project team members to play the role of a quality manager after providing him/her with training.
>
> Having determined measurements for the project quality requirements doesn't mean that you have to wait until the task ends to assess the accuracy of the measure. For example, we identified that the meat should be cooked until it's well done, so this should prompt us to periodically check on doneness while cooking. This is called quality audit, in which we assess the task for the quality of its output even before it is finished, and if we expect any possible deviation from the quality measure, we should try to amend it before it is too late. Quality audits can be scheduled in the project time plan at predetermined dates or in ad hoc periods.

**Gold Plating**

By "gold plating" we mean putting in place EXTRA quality requirements that will not add benefits to the final output of the project, as in the following examples:

| Project | Example of Gold Plating |
|---|---|
| Building a residential high-rise | Install earthquake harmonizing systems in an area not known for earthquakes |
| Building a laptop | Building the cover out of titanium |
| Creating a website | Buying much more storage area than needed |

Note in the second example that building a laptop out of titanium might not be gold plating if it will be used in tough terrains and needs to be as durable as possible.

**REEM Park Example: Quality Plan**

Amanda and the project team filled in the following template :

Template 11

*Project Quality Requirements*

| No. | Task | Required Quality | How to Measure It |
|---|---|---|---|
| 1 | Hiring a consultant company | Select the best company based on proposals | By using a clear method to assess and select from among the different companies |
| 2 | Park design | Design should be free of error | Review of the design will be done by different members of the park department |
| 3 | Hiring a contractor | Select the best company based on proposals | By using a clear method to assess and select from among the different companies |
| 4 | Construction | Construction as per international standards<br><br>Children's play equipment to resist weather conditions for a period of 10-15 years | By conducting frequent quality audits<br><br>Check the specifications of the equipment before it is approved |

### Benefits of Using This Template

1 You reviewed the project scope and project map templates.

2 You thought about the required quality for your project.

3 The quality template aided you in identifying risks during risk management.

# Time Plan

Estimating the duration of your project with great detail and accuracy is a very challenging task which can, nevertheless, be enjoyable. Challenges arise from the existence of different tasks in the project that need to be estimated, while the joy comes from the opportunity to use different project management tools to think about and analyze your project. The tools we will be using in this section are the Gantt Chart and the network diagrams.

### Benefits of a Project Time Plan (Schedule)

1 How long is your project going to last? This might be the most asked question about your project. With a proper time plan, you will be able to answer it with confidence.

2 You will be able to know when individual tasks in your project will start and end, which will then enable you to make informed decisions (like moving tasks back and forth).

3 Delayed projects are very common in project management. With a project schedule, you will be able to compare actual progress with what was planned, and if you discover a delay, you can act early to fix it.

### How to Make a Project Time Plan

To make a time plan, we need to define two things: how long each task will take and the link between tasks (that is, which tasks precede the others). We need to conduct the following activities:

1 Review the project map.

2 Estimate the duration of each individual task in the project.

3 Determine the link between the different tasks;4 Draw a time plan for the project.

## First: Review the Project Map

You start with this to avoid any reworking that might be needed because of missing some tasks. For example, in the dinner project, we might see the need to add a new task called "prepare and serve" which consists of preparing the food. So the new project map will look like this:

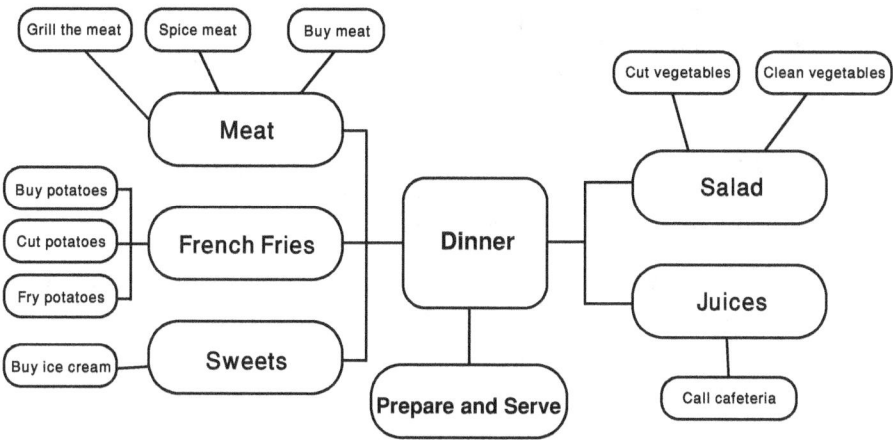

## Second: Estimating the Duration of the Project Tasks

Now we will use the project map to list all the tasks we identified and then try to estimate their duration. There are different methods for estimating the duration of each task, but the most intuitive and straightforward are:

- Estimation by asking experts;

- Estimation by comparison with similar tasks;

- Using the internet (which can cover the earlier methods of asking experts and comparison with similar projects).

Let's take the example of the dinner project, where, if you want to estimate the time required to marinate the meat, you can simply ask one of your friends who is an "expert" in cooking. In addition, to estimate the duration of time required for the cafeteria to deliver the juice, you might recall that in the past it took an average of 30 minutes, so you can use this duration for your project. When estimating the duration of the tasks, you should be able to justify why you used a particular number. It would also be a good idea to record the sources of your information so that you can review them when conducting risk management to examine whether they are reasonable and unbiased. After estimating the duration of your project tasks, just note them down on the project map as illustrated in the dinner example:

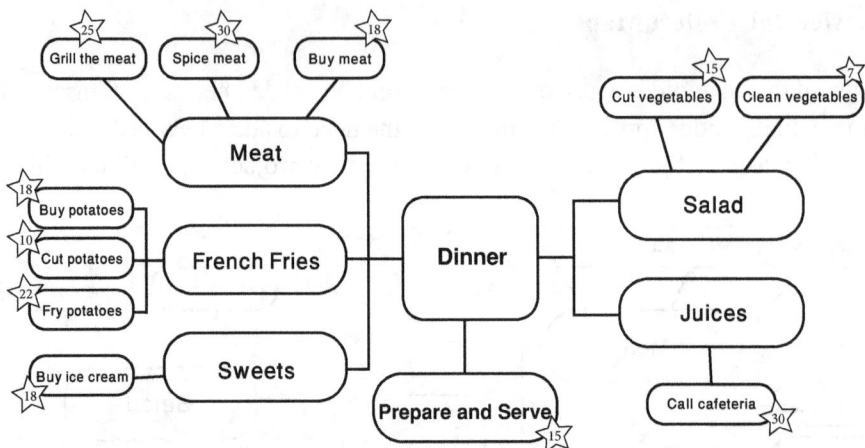

Note that if you face difficulties in assigning time to a particular task, it might be because you need to break it down into subtasks.

### Third: Arrange Tasks in a Sequence

Now we will link the different tasks together in a time sequence, going from the beginning of the project to its end. In the simplest case, the project will have only one path from beginning to end where no two tasks will happen simultaneously, as follows:

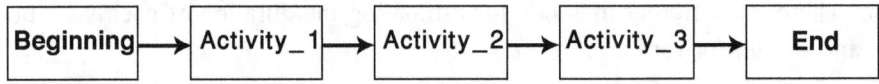

In other cases, there will be different paths in a single project wherein some tasks can happen at the same time as others, as illustrated here:

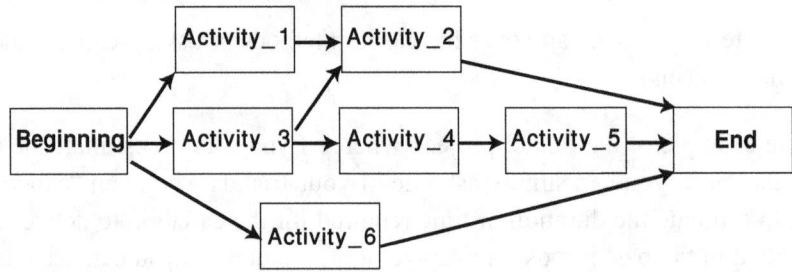

For the above diagram, note the following:

• The diagram is called a "network diagram."

• Every way of getting from the start of the project to the end is called a "path."

- Tasks are represented in squares and the arrows describe the "precedence relationship" between them (where no task can start until the one before it is finished).

- Note that some tasks don't depend on others in order to start, while others depend on one or more tasks having been completed.

**Exercise: Make a Network Diagram for the Dinner Project**

To do this exercise, we need a copy of the project map with the duration of each task noted on it. Then, we will use a blank paper, draw a square on the left and write "beginning" in it and a square on the right and write "end" in it. Now, we start putting the project tasks in between the two squares by asking the following two questions for each task:

1 What task precedes this one?

2 What task follows this one?

Note that this exercise will involve some trial and error, so don't be disappointed if you don't get it right the first time (be prepared to use many pieces of scratch paper!).

After you finish with your network diagram, double-check with the project map to make sure you did not omit any task.

**The following is a network diagram for the dinner project:**

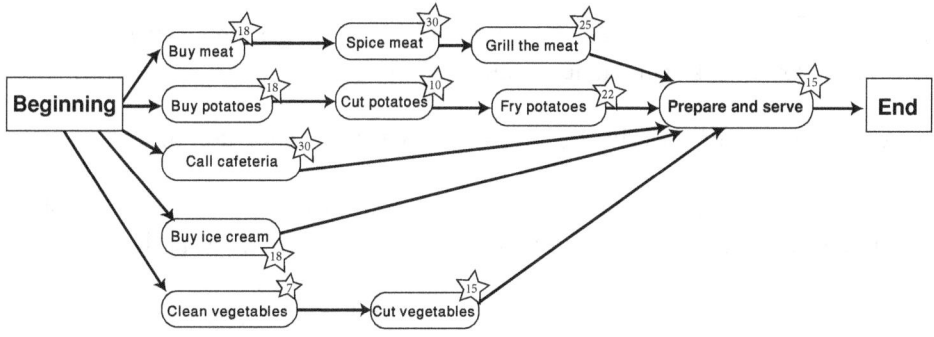

> When making the network diagram, we assume the availability of resources because this will help us determine the optimal paths for the project. To illustrate what I mean, consider the dinner project example wherein we placed "cutting potatoes" and "cutting vegetables" as two simultaneous tasks. But in order for them to happen at the same time, you need adequate resources (that is, if you have only one person working, he/ she cannot perform both tasks at the same time). So, as a rule of thumb when doing the network diagram, always treat the project as if you have unlimited resources. Once we actually calculate the resources for the project, we can fix the diagram if we discover we don't have sufficient resources to perform certain tasks at the same time.

> Constructing a network diagram will help a project manager achieve the following strategic goals:
>
> 1 Harnessing opportunities: It might be possible to do some tasks at the same time, thus reducing the overall duration of the project.
>
> 2 Avoid problems: Since some tasks cannot be started until others have been completed, ignoring or forgetting this step can delay the project.

## Fourth: Draw Out the Project Time Plan (Schedule)

Here we are going to convert the network diagram into what is known as a Gantt Chart or, simply, the project schedule. This schedule will be the baseline for the project, meaning that the project manager can print it on a big piece of paper and post it in his/her office to track the actual project progress with it. To draw the time plan, we need the network diagram, a ruler and some colored pens.

## Exercise: Make a Time Plan for the Dinner Project

To make the time plan, you will need to use the network diagram.

1   Start by writing the tasks down the left side of a piece of paper.

2   Choose the unit of time and write it on top of the page (in the dinner example, we used minutes). Also, write the starting date (time).

3   Divide the page into columns, where each column will represent a unit of time.

4   Represent each task with a line whose length corresponds to the duration of the task. So if the task will take 5 minutes, the line will take the length of 5 columns as shown below:

|       | 1 | 2 | 3 | 4 | 5 | 6 | 7 |
|-------|---|---|---|---|---|---|---|
| Task1 |   |   |   |   |   |   |   |

The following table represents the time plan for the dinner project:

Start Time = 7 pm

| No. | Task | Time | 10 | 20 | 30 | 40 | 50 | 60 | 70 | 80 | 90 | 100 |
|---|---|---|---|---|---|---|---|---|---|---|---|---|
| 1 | Buy meat | 18 | ▆▆▆▆ | | 18 | | | | | | | |
| 2 | Spice meat | 30 | | 18 ▆▆▆▆ | | | 48 | | | | | |
| 3 | Grill the meat | 25 | | | | | 48 ▆▆▆▆ | | | 73 | | |
| 4 | Buy potatoes | 18 | ▆▆▆▆ | | 18 | | | | | | | |
| 5 | Cut potatoes | 10 | | 18 ▆▆ | 28 | | | | | | | |
| 6 | Fry potatoes | 22 | | | 28 ▆▆▆ | | 50 | | | | | |
| 7 | Buy ice cream | 18 | ▆▆▆▆ | | 18 | | | | | | | |
| 8 | Clean vegetables | 7 | ▆ 7 | | | | | | | | | |
| 9 | Cut vegetables | 15 | 7 ▆▆ | 22 | | | | | | | | |
| 10 | Call cafeteria | 30 | ▆▆▆▆ | | | 30 | | | | | | |
| 11 | Prepare and Save | 15 | | | | | | | | 73 ▆▆▆ | 88 | |

Time unit = minutes  —  Start Date / End Date

**Comment 1:** Obviously, we will not start slicing the potatoes immediately after buying them, but for the sake of simplification, we did not add any waiting time (because we will start preparing the other items, like the meat and vegetables, at the same time as the potatoes, so we can discard the waiting time).

**Comment 2:** The time plan we make in this section will not be considered complete because we haven't considered risks. In the next chapter, we will apply risk management to our time plan and finalize it.

**REEM Park Example: Time Plan**

Amanda followed the following sequence to produce a time plan for the park:

**First:** A meeting was conducted with the project team and experienced staff from the parks department. During the meeting, the project map was reviewed and the duration of each task was estimated. The following project map shows the estimated duration for each activity:

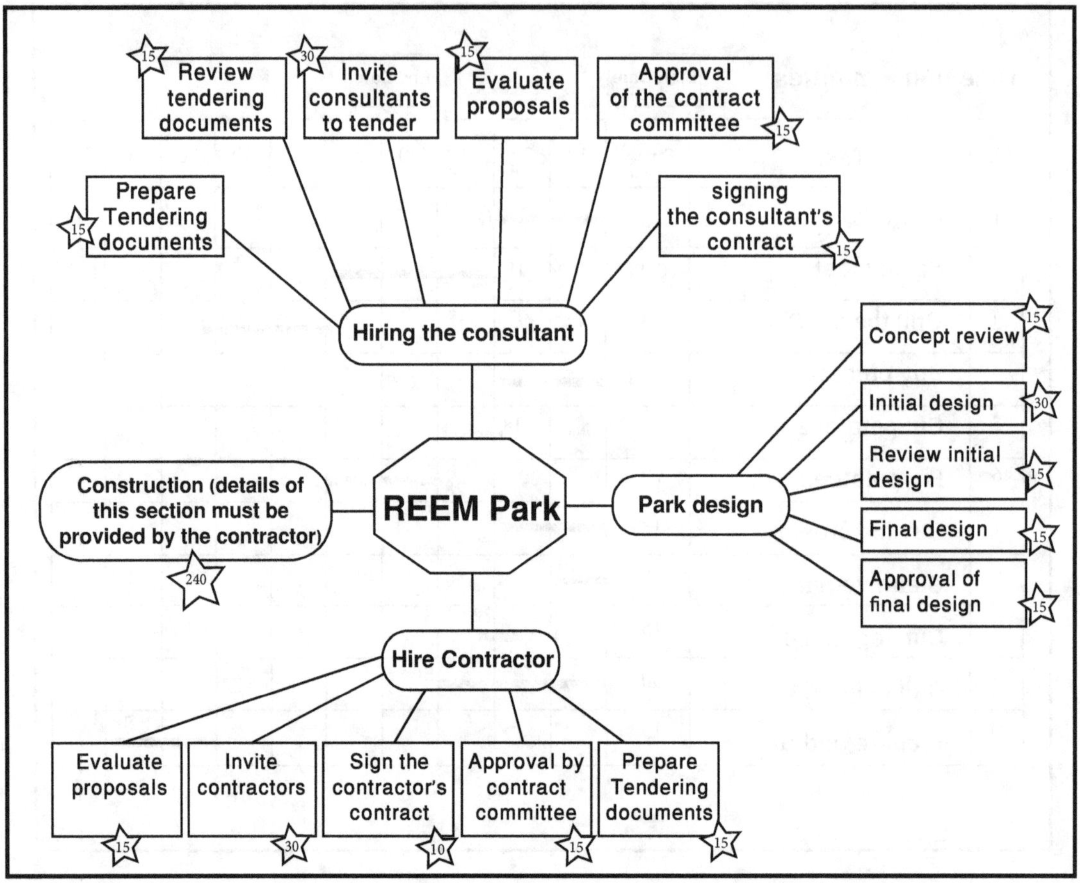

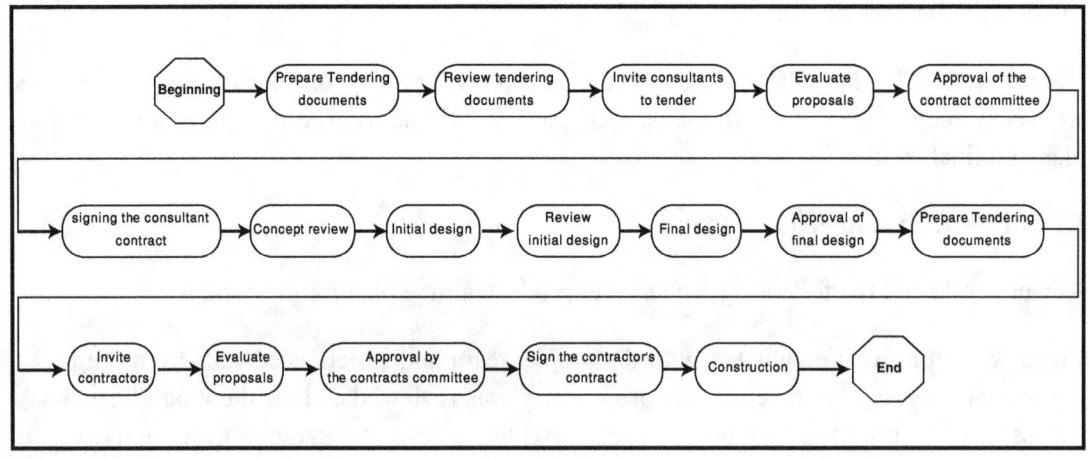

Review tendering documents
15
Invite consultants to tender
30
Evaluate proposals
15
Approval of the contract committee
15
Prepare Tendering documents
15
signing the consultant's contract
15
Hiring the consultant
Concept review
15
Initial design
30
Review initial design
15
Final design
15
Approval of final design
15
Park design
REEM Park
Construction (details of this section must be provided by the contractor)
240
Hire Contractor
Evaluate proposals
15
Invite contractors
30
Sign the contractor's contract
10
Approval by contract committee
15
Prepare Tendering documents
15

**Second:** A network diagram was done for the project as follows:

Beginning
Prepare Tendering documents
Review tendering documents
Invite consultants to tender
Evaluate proposals
Approval of the contract committee
signing the consultant contract
Concept review
Initial design
Review initial design
Final design
Approval of final design
Prepare Tendering documents
Invite contractors
Evaluate proposals
Approval by the contracts committee
Sign the contractor's contract
Construction
End

**Third:** The two works above were combined to produce the time plan for REEM park as follows:

Template 12

Time unit is month. Start date is 1 July 2010.

### *Project Time Plan*

| No. | Task | Time | 1 | 2 | 3 | 4 | 5 | 6 | 7 | 8 | 9 | 10 | 11 | 12 | 13 | 14 | 15 | 16 | 17 | 18 |
|---|---|---|---|---|---|---|---|---|---|---|---|---|---|---|---|---|---|---|---|---|
| 1 | Prepare tendering documents | 15 | ■ | | | | | | | | | | | | | | | | | |
| 2 | Review tendering documents | 15 | ■ | | | | | | | | | | | | | | | | | |
| 3 | Invite consultants to tender | 30 | | ■ | | | | | | | | | | | | | | | | |
| 4 | Evaluate proposals | 15 | | | ■ | | | | | | | | | | | | | | | |
| 5 | Approval of the contract committee | 15 | | | | ■ | | | | | | | | | | | | | | |
| 6 | Signing the consultant contract | 15 | | | | ■ | | | | | | | | | | | | | | |
| 7 | Concept review | 15 | | | | | ■ | | | | | | | | | | | | | |
| 8 | Initial design | 30 | | | | | ■ | | | | | | | | | | | | | |
| 9 | Review initial design | 15 | | | | | | ■ | | | | | | | | | | | | |
| 10 | Final design | 15 | | | | | | | ■ | | | | | | | | | | | |
| 11 | Approval of final design | 15 | | | | | | | ■ | | | | | | | | | | | |
| 12 | Prepare tendering documents | 15 | | | | | | | | ■ | | | | | | | | | | |
| 13 | Invite contractors | 30 | | | | | | | | ■ | | | | | | | | | | |
| 14 | Evaluate proposals | 15 | | | | | | | | | ■ | | | | | | | | | |
| 15 | Approval by the contracts committee | 15 | | | | | | | | | ■ | | | | | | | | | |
| 16 | Sign the contractor's contract | 10 | | | | | | | | | ■ | | | | | | | | | |
| 17 | Construction | 240 | | | | | | | | | ■ | ■ | ■ | ■ | ■ | ■ | ■ | ■ | ■ | ■ |

**Note:** Weekends and holidays were incorporated into the plan, so if a task takes one month, it includes all holidays. This is done for simplification; else, we can separate holiday, and even rest hours, from the project work.

**Benefits of Using This Template**

1 You reviewed the project map.

2 You estimated the duration of your project tasks.

3 You organized your project tasks in a sequence.

4 You have a diagram that summarizes your project in one or a few pages.

# Project Monetary Plan

Our goal here is to come up with one figure describing how much the project will cost. To determine this, we need to first identify the resources needed for each task and then determine their cost.

The following list shows examples of resource requirements:

- People (either laborers or specialized or skilled workers);
- Materials;
- Testing and analysis (such as testing samples of concrete in a construction project);
- Equipment (such as cars and computers);
- Consumables (such as fuel and utilities);
- Space (such as offices).

**How to Estimate Cost**

Similar to estimating the time periods, we can use the following methods to estimate cost:

- Estimation by asking experts;
- Estimation by comparing the tasks to similar ones;
- Using the internet.

For example, if you are installing wooden floors in your house as part of a renovation project, you can estimate the cost by calling some shops and getting an approximation of how much it will cost to cover the area that you have. Also, if a company is installing network cables in a particular floor, it can find the cost by reviewing similar projects conducted for other floors.

**Exercise: Estimate the Cost of the Dinner Project (Determine the Required Budget)**

First, we will create a resources table wherein we will:

1 Identify the resources needed for each task (we assume that the dinner is planned for five people);

2 Identify whether resources are available or if we have to buy or rent them;

3 Estimate the cost of the resources;

4 Add up all the costs to create an initial project budget (the budget will only be finalized after risk management).

---

**Note 1:** The resource table will summarize many processes you will have used in order to identify the needed resources and estimate their cost. You should save the scratch papers you have used for the calculations so you can review them during risk management.

**Note 2:** In some projects, the cost of certain resources might be small enough that we can just ignore them, such as the cost of fuel used to go shopping for food in the dinner project. On the other hand, the cost of fuel might be significant (in the thousands of dollars) for a construction project wherein many trucks are used to transport materials.

**Note 3:** When we made the network diagram, we assumed we had plenty of resources. But when we actually set the budget, if we discover that we don't have enough resources to perform two or more tasks at the same time, we will have to amend the network diagram by putting such tasks in sequence instead of parallel to each other. Remember that the time plan depends heavily on the network diagram, so if we change the network diagram we will need to revise the time plan

---

Template 13

*The following is the resource table for the dinner project:*

| No. | Task | Human Resources | Cost/Comments | Equipment/ Materials | Cost/ Comments |
|-----|------|-----------------|---------------|---------------------|----------------|
| 1 | Buy meat | One person | By project manager | Car + Cost of meat | $25 for meat |
| 2 | Spice meat | Will use a worker from a nearby restaurant | $20 | Spices | Available |
| 3 | Grill meat | One person | By project manager | Grill | Available |
| 4 | Buy potatoes | One person | By project manager | Cost of potatoes | $7 |

| 5 | Cut potatoes | One person | By daughter | Knife | Available |
|---|---|---|---|---|---|
| 6 | Fry potatoes | One person | By daughter | Cooking oil | $8 |
| 7 | Buy ice cream | One person | By project manager | Cost of ice cream | $7 |
| 8 | Wash vegetables | One person | By wife | Vegetables | $5 |
| 9 | Cut vegetables | One person | By wife | Knife | Available |
| 10 | Call restaurant for juice | One person | By project manager | Cost of juice + Phone | $22 for juice |
| 11 | Prepare and serve | One person | By project manager | Dishes | Available |
| Total | | $20 | | $74 | |
| Project Budget | $94 for the cost of dinner (for 5 people) | | | | |

Important: You might discover that you need additional resources that will require you to amend your project time plan. For example, in the dinner project, we identified the need for cooking oil, so we need to add into the project map the task of purchasing oil. Subsequently, we will need to add it into the project network diagram and time plan.

### REEM Park Example: Project Budget

The REEM project is different because the cost will be the amount of money we will pay for the consulting and contracting companies, which would typically be paid in installments. Before calculating the budget, I need to explain the following point:

- **The Difficulty of Calculating Costs in Some Projects**

Estimating the cost of project tasks becomes difficult as the project size and complexity increases. Take the example of determining the cost of installing electrical wiring in a house construction project. We can find the cost by using the following formula:

Cost of a Task = (types of equipment needed * number of each type * cost of each type) + (types of materials needed * number of each type * price of each type) + (type of specialities of workers* number of each type * wage of every worker)

Note that calculating such quantities will require a lot of effort searching for information and making calculations; as a result, many companies employ people to make such calculations.

In the case of the REEM project, the consultant and contractor will provide the cost for the project, but Amanda must make sure that the cost is fair. She also needs to provide the municipality (that is, the finance department) with a figure to be used as a basis for setting aside a "money reserve" for the project. What Amanda can do is to determine the fair market price by contacting several consulting and contracting companies. For parks and construction projects, the price is usually calculated based on each square meter (or square foot). So we assume that Amanda called a few companies and found the average market price for one square foot to be $25. Now all she needs to do is multiply this by the area of the park:

So, cost of the park = area in square feet * price per square foot

$$= (200 * 400) * 25 = \$2,000,000 \text{ million US.}$$

Now Amanda and the project team will have an estimated budget they can provide the finance department and use as a basis against which to compare the cost that will be provided by the consultant and contractor.

**Note**: The estimated cost of $2 million is inclusive of both the consultancy fee (to design the park) ad the contractor's cost (to construct the park). In the business of parks construction, the consultant usually gets a percentage of the construction cost (such as 4%). Also, payments are given in predetermined installments decided by the sponsoring companies (for example, the consultant might be paid 20% of the price after signing the contract and 50% after providing the initial design). Similarly, the contractor might be paid in installments corresponding to the progress of construction. Every business has a different way of calculating costs for a project, and the project manager must always follow the organization's standard procedure for estimating the costs of his/her project.

### Benefits of Using This Template (Resource Table)

1 You reviewed the project map.

2 You estimated the resources required for the project tasks.

3 You estimated the cost of different tasks.

4 You calculated the estimated budget for the project.

# Contractors Selection Plan

In many cases, the project team will not be able to complete the project by themselves and will require external assistance. These are called contractual services, such as the need for a company to design and construct a high-rise building or for a company to create a website or for a company to do a study on the effectiveness of the internal procedures of an organization. But how do you select the best company among different candidates? To make the best selection, we need to first establish criteria upon which we will evaluate different companies.

**The Process of Selecting Contractors:**

Usually, selecting external help for your project will pass through the following steps:

1  From the project map, you will identify the need to hire a contractor of some kind.

2  Prepare documents explaining your needs and what service/s you require, including conditions.

3  Send these documents describing your needs to the potential contractors who can satisfy them.

4  Receive documents from the interested contracting companies explaining their price and how they will do the work.

5  Evaluate the proposals and select a contracting company.

6  Sign a contract with the selected company and track its performance as it provides the required service.

All of the above steps will take place during project execution (except for the first one, of course). However, we are now in the project planning stage, and our job is to prepare for project execution. Therefore, we will now prepare the selection template. To prepare the template, we need to determine the following:

**First: The Balance of Price and Quality for Service Provided**

When a company submits a proposal, it will provide two kinds of information: its price for conducting the service (called the financial offer) and the method and means by which it will provide the required service (called the technical offer). If we evaluate a proposal based on only one kind of information, we will risk exposing the project to the following risks:

| Situation | Risk |
|---|---|
| The contractor company is selected based on the price only | Some companies might sacrifice quality to provide a competitive price<br>Big companies might not be interested in participating in the project tender as they cannot accept lower profit margins like smaller companies can |
| Selection is based only on the quality of work | Prices might be exaggerated<br>Smaller companies will not be able to compete with larger ones which often offer better quality work |

Therefore, we need to balance the price and quality of a service (like we do in our everyday life, as in choosing between buying products and services with different prices and options). We can provide a percentage for the financial and technical offer to balance both, as shown in the table below:

| Intention | Example of Percentages |
|---|---|
| If a company is after the best quality | Financial offer  20%<br>Technical offer 80% |
| If a company is after the best prices | Financial offer  10%<br>Technical offer 90% |
| Balance between price and quality | Financial offer  50%<br>Technical offer  50% |

Note that figures presented in the table are just for example, and each company will have its own balance between cost and quality, depending largely on the importance of the project and availability of funding.

## Second:  Deciding on the Evaluation Criteria

Evaluation criteria are the elements that will reveal to us to what degree the services of a contracting company are suitable for our project.  The criteria must include various elements so as to give the project team a well-rounded idea of a contracting company.  For example, a company might be very good in giving the service we need but is not experienced in this part of the world, which may increase the chances of it facing certain difficulties, such as working with local suppliers and authorities.

For the evaluation of the financial offer, there will usually be only one criterion: how much a company will charge.  As for the evaluation of the technical offer, we will likely have different elements to consider, such as:

- The qualification of the project team from the contracting company;

- Experience of the company;

- Whether the company has proper certification/s from international systems, such as ISO 9001.

**Third: Deciding How to Balance the Evaluation Criteria**

After deciding on the elements to be used in evaluating the technical offers, we need to decide how to balance them. This will depend on the type of service needed. For example, if the service is to provide an intellectual product, we might give more weight to the academic qualifications of the people providing the service.

**REEM Park Example: Contractors Evaluation Plan**

First of all, Amanda contacted the contracts department to identify the official procedure for balancing the financial and technical offers, which she found to be 50% each. However, Amanda wasn't satisfied with this distribution as it would increase the chance of selecting contractors with less than adequate quality work (but offering a lower price). As a result, she met with the manager of the parks department and convinced him to write a letter to the manager of the contracts department explaining the importance of REEM Project and some past cases with parks projects that encountered difficulties and were delayed due to bad contractors. In the letter, the manager of the parks department asked to change the percentage to 30% for the financial offer and 70% for the technical offer, which should increase the chance of selecting better quality contractors rather than cheaply priced ones.

After that, Amanda assembled the project team and brainstormed about what elements should be included in the technical offer evaluation. The team agreed to include the following elements:

- Experience of the company in the technical field of the project;

- Similar projects conducted;

- Experience of the company in the region;

- Qualifications of the project team members;

- Compliance with the scope of the service provided by the REEM project team;

- Possessing an ISO 9001 certification.

Amanda also suggested the additional element of assessing the financial stability of the company because of the financial crisis that had hit the global market. This can be done by asking the company to submit a financial statement and some bank guarantees.

After that, the project team decided on the percentages for the various evaluation elements and prepared the evaluation template as follows:

Template 14

*Technical Evaluation for REEM Park*

| Comparison Criteria | Weight | Company 1 | Company 2 | Company 3 | Company 4 | Company 5 |
|---|---|---|---|---|---|---|
| • Experience of the company in the technical field of the project | 10 | | | | | |
| • Similar projects conducted | 10 | | | | | |
| • Experience of the company in the region | 10 | | | | | |
| • Qualifications of the project team members | 25 | | | | | |
| • Financial stability | 10 | | | | | |
| • Compliance with the scope of the service prvided by the REEM project team | 25 | | | | | |
| • Possessing an ISO 9001 certification | 10 | | | | | |
| Total | 100 | | | | | |
| Signature | | | | | | |

Comment: The contractors evaluation plan, as was discussed in this section, applies to medium to large projects wherein the description of the required service is written and officially sent to potential contractors. However, the same principles can be used for smaller projects for which you can prepare the selection criteria in advance and fill in the template by interviewing potential contractors onsite or via phone calls.

**Benefits of Using This Template**

1   You made a template that will guide you in evaluating different proposals from companies to work for your project.

2   This template will help in documenting the evaluation process, which is important from a legal point of view.

# Project Change Request Plan

You will not be able to entirely eliminate changes to your project, and it may happen that some changes prove beneficial to the project. The project must be flexible; instead of overlooking change requests, it is better to design a system for reviewing such requests and responding to them in a timely and organized manner. The following are some examples of possible changes in a project:

•   The sponsor requests the addition of one more room to his house;

•   The IT department wants to add a new module to allow voice-mailing in its new activity-tracking project;

•   Because the price of marble went down, the sponsor requests changing the flooring of the dining room in his house to marble instead of wood;

•   The contractor requests from the sponsor an extension of time for various reasons.

To organize the process of making changes, the project manager needs to do the following:

1   Prepare a change request template and make it available for stakeholders who are influential enough to effect change in the project;

2   Establish a mechanism to respond to and evaluate change requests.

---

**Note**: Some change requests might require immediate response, even before filling out a change request form. Take the example of an accident that happens at a construction site that demands an immediate change in work methods.

---

**REEM Park Example: Change Requests**

Amanda and her team put in place the following change request form:

Template 15

| Change Request Form No.: REEM Park Project | |
|---|---|
| Change Description | |
| Reasons for Change | |
| Name of Person Submitting | |
| Date | |
| Possible Effects if the request is accepted | |
| Is the Request Approved? | YES    NO<br>Reasons: |
| Signature of the Project Manager | |

The team also created the following system to respond to change requests:

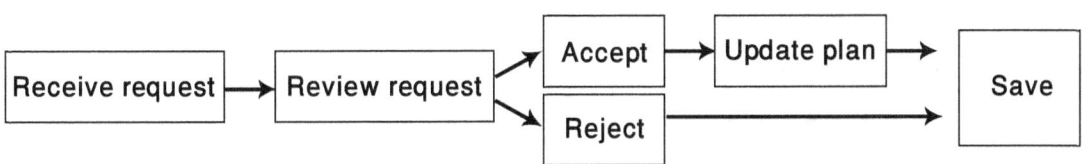

**Benefits of Using This Template**

1  By using this template, you are organizing the process of change in your project.

2  The template will help to document all change requests.

3  By having a template to fill out and explain reasons for the change, people will think twice before asking for any change.

# Project Progress Reporting

During project execution, the project manager should issue progress reports that contain information such as:

- Percentage of work completed against what was planned;
- Risks that the project has encountered or is facing;
- Lessons learned;
- Upcoming tasks;
- Financial information.

This information can be reflected in a template to facilitate and expedite its issuance.

**REEM Park Example: Progress Report Template**

The project team agreed that Amanda will issue a monthly progress report during the execution stage and prepared the following progress report template:

Template 16

### *REEM Park Project Progress Report*

| Report Number ......<br>Date ........ | |
|---|---|
| Project Status | |
| Days of Delay | |
| Tasks Completed This Month | |
| Risks Facing the Project | |
| New Risks | |
| Upcoming Tasks | |
| Lessons Learned | |
| Project Manager's Signature | |

**Benefits of UsingThis Template**

You identified the information you would like to share about your project during progress reporting.

# Chapter Five
# Final Planning

In this chapter, we will use the tools and techniques of risk management to review our planning, represented in the templates we have made so far. Project execution will be happening quite soon, and the project team must make sure the templates used for planning contain minimal errors and contradiction among them so they can be used with confidence during project execution.

# Risk Management

Identifying risks and putting measures in place to control them might be the single most important thing a project manager can do for his/her project. By the word "risks," I mean events that can arise during project execution and negatively affect its success, such as delaying the project, increasing its budget requirements or even killing it. Based on this definition, be sure you understand that risks are possible future events that may or may not happen. Regardless, we decide to take action (that often consumes some resources) because we feel it's better to be prepared in case such events arise. The following are some examples of risks in projects:

- The delivery of equipment required for certain project tasks is delayed;

- A sudden increase in the price of materials;

- The contracting company for the project files for bankruptcy;

- Poor communication between the project team and major stakeholders;

- Problems with providing the required cash payment/s for the project;

- New governmental or international laws that affect the project;

- Mistakes in designs, specifications, etc.;

- The death of the project manager!!

**Benefits of Risk Management in Projects**

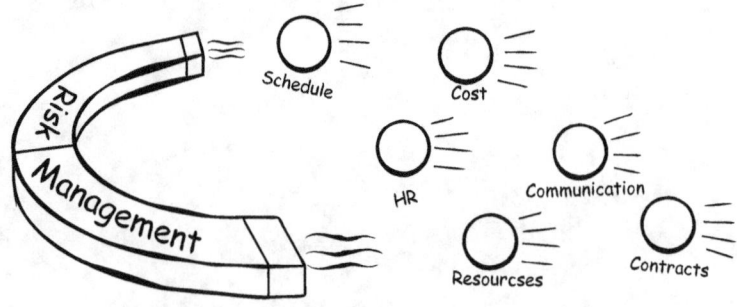

Risk Management pulls everything together

I like to use the analogy that risk management works like a magnet because it pulls together the different elements of a project plan (like stakeholders management, time plan and quality plan) so you can take a better look at the project and identify possible conflicts, such as:

- The quality requirements for a certain task require more money than what was budgeted;

- The information sharing plan fails to include important stakeholders;

- Available resources don't allow for certain tasks to be done simultaneously as indicated in the time plan;

Risk management provides the project team with the opportunity to review their planning and acts as a test to all assumptions made.


## Project Sensitivity to Risks

Conducting risk management requires a great deal of time and resources, so before you start, you should determine how thorough you will be in your analysis and how many resources you will be willing to use to control risks. The rule of thumb is that the more important a project is, the more sensitive we should be in dealing with risks. Being more sensitive means that the project team will spend more time identifying risks and will try to respond to a larger number of risks. Also, the team will be willing to allocate more of the project budget to control risks. The following are examples of what might increase project importance, leading to the increase of sensitivity to risks:

- The project is directly linked to achieving organizational strategic goals;

- The project will affect the reputation of the organization;

- A large amount of money is being invested;

- The time plan for the project is fixed with no possibility of extension.

Determining risk sensitivity can be done in a quantitative or qualitative way. The qualitative way is easier and faster. First, the project manager should discuss with the sponsor the importance of the project on hand and how many resources should be allocated to prevent any problems in achieving it. The project manager should then brainstorm with the project team to give the project a score representing its level of sensitivity (importance): high, medium or low. The following are examples of how to use risk sensitivity:

1  If risk sensitivity is low and there are differences of opinion in ranking a risk, the lower ranking may be adopted.

2  If sensitivity is high, the team should be more thorough in identifying risks.

3  If sensitivity is high, the team will be willing to use more resources to control risks.

## Life Cycle of Risk Management

When conducting risk management, we will go through the following stages:

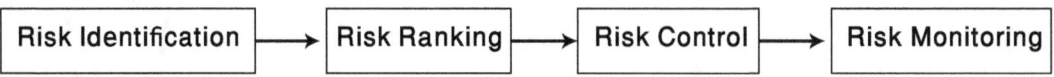

## Stage One:  Risk Identification

Identifying risks is an exercise of intelligently guessing what might go wrong in your project in the future (that is, during its execution).  Therefore, you should think of all different possibilities and scenarios that the project might face that would result in a negative outcome.  There are many ways of identifying risks; I will present two of the most useful:

- **First: Brainstorming regarding the Project Scope**

The project manager holds open meetings with the project team and selected stakeholders to examine the project scope and think openly about what might go wrong in it.  Similar to any brainstorming session, you should encourage participation and not judge the quality of ideas.  Also, these sessions can be improved by inviting experts in the technical field of the project who can identify possible risks based on their experience.  So, if you have an IT project that involves programming, you might find it useful to invite a programmer to examine the project scope and give his/her thoughts about what might go wrong.

Sometimes the project manager and the project team will be new to the organization sponsoring a project.  In this case, it will be very important to invite people who have worked in the organization for a while to help determine risks that might emerge from internal procedures (like difficulties securing approvals or some lengthy procedures).

- **Second: Reviewing Project Documents**

By now we have accumulated many templates and, thus, can review them for completeness and errors. Also, we can cross-check them for conflicts. The following table suggests a method of reviewing templates:

| Template | How to Review It |
|---|---|
| Feasibility Study | Compare the original project goals with the scope template for consistency |
| Project Team Plan | Are responsibilities distributed fairly?<br>Are there conflicts between responsibilities?<br>Is the workload reasonable?<br>Are all tasks assigned to someone?<br>Did all the team members sign onto their responsibilities? |
| Stakeholder Management Plan | Did we identify all stakeholders?<br>Are there conflicts between stakeholder requirements and the project objectives? |
| Information Sharing Plan (communication) | Did we identify all people requiring information about the project?<br>Is the means and frequency for sending information adequate?<br>Did we identify the person from the project team responsible for sending information? |
| Project Map | Did we identify all tasks in the project? |
| Quality Plan | Is it reasonable (and possible to attain the required quality) in light of the available time and resources?<br>Does the quality include Gold Plating? |
| Time Plan | Did we accurately estimate the time needed to complete tasks?<br>Did we arrange the tasks correctly in the network diagram? |
| Budget | Did we accurately estimate the resources (and money) needed to complete different tasks?<br>Do we have enough resources to comply with the time plan? |
| Contractors Selection Plan | Are the selection criteria appropriate for the project type?<br>How thorough and diverse are the selection criteria? |
| Project Change Plan | Do we have a change request template?<br>How will we make the template available to stakeholders?<br>Are roles and responsibilities clear in regard to receiving and reviewing the change templates? |

To review the templates, I suggest following these steps:

1 Collect all the templates done for the project so far.

2 Make copies and distribute them to the project team.

3 Try to distribute copies of the templates to people who did not participate in filling them out (such as your colleagues at work and experts in your company).

4 Ask everyone to review the templates (using the table above as a guide) and to attend a meeting the next day to discuss their observations.

5 Collect all concerns as potential risks and add them to the risks collected using the first method.

---

**This step of reviewing the project templates is very important for the Project Book, so make sure you don't skip it.**

---

**Exercise: Find the Potential Risks in the Dinner Project**

The following is a list of potential risks:

- Not finding good quality meat;

- Sudden increase in food prices;

- The car breaks down;

- The meat gets burned during cooking;

- Not finding good quality potatoes in the market;

- The gas cylinder goes empty during cooking;

- The cafeteria is delayed in delivering the juice;

- The potatoes are undercooked.

# Stage Two: Risk Ranking

Ranking risks is important for two reasons:

1  The quantity of risks: Chances are you will identify a lot of potential risks through brainstorming and reviewing the project templates, and responding to all of them will be considered a waste of time because not all of them will be serious.

2  Resources are limited: Your goal in risk management is not just to identify risks, but to control them. However, controlling risks will usually require you to use some of the allocated project resources (like people, time and money). Some risks will not have high potential of happening during project execution, and even if they do happen, they will have a minimal effect that will not justify spending project resources.

Risk ranking is a logical and straightforward exercise, but to do it smoothly, we need to understand the following two points:

**First: The Risk Formula**

We can represent (that is, define) risks mathematically with the following formula:

**risk = chance of occurrence * potential effect**, where:

risk = an event that will cause loss to the project (an example of risk is delay);

chance of occurrence = the likelihood that the event will actually happen in the future (an example is that the chance is highly likely); and

potential effect = if the event happens, how strong of an effect might it have on the project? (example of effect is a one-day delay).

This formula is very important because we will actively use it in ranking the risks and, later, in controlling them. It is powerful because it enables us to think of the basic components of risks (chance and effect) and understand them better, which in turn will help us make better decisions based on clear understanding and not on guessing. From the risk formula, we can infer the following:

•  If the chance of occurrence increases (or decreases), then the risk will increase (or decrease) accordingly;

•  If the potential effect increases (or decreases), then the risk will increase (or decrease) accordingly.

So, we will depend on BOTH the chance and effect for ranking risks. The following table gives examples of the elements of the risk formula:

| Risk (Event) | Chance of Occurrence | Potential Effect |
|---|---|---|
| There are not sufficient funds to start executing the project | LOW because communication with the finance department was established beforehand | HIGH because it can lead to canceling the project |
| Delay of required equipment | MEDIUM because the equipment has to come from a neighboring state | MEDIUM because it is possible to rent the equipment |
| Sudden increase in wood prices during a project to construct a house in the countryside | LOW because there is no indication in the wood market of such possibilities | HIGH because a lot of wood will be needed for the house |
| Lack of communication between the project team and some important stakeholders | HIGH because they were not identified | HIGH because these stakeholders possess the strength and interest to sway the project toward their needs |

**Second: The Risk Matrix**

Because a risk is a product of two factors (chance and effect), we can use a matrix to find the result faster. We will call this a risk matrix, which is basically a table where chance is represented vertically and effect horizontally (or vice versa), as in the figure below:

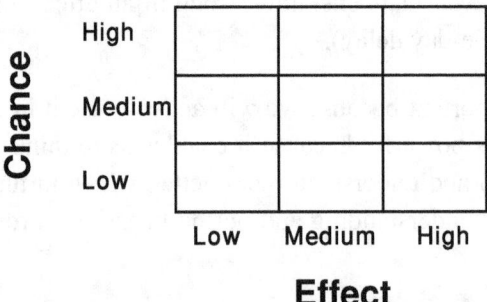

Note from the table that there are three possible scores for chance and three possible scores for the effect: low, medium and high. So there are nine possible results, illustrated by the small squares which represent the magnitude or score of the risk in question.

Using the risk matrix is very easy: Just determine the chance and effect of the risk and draw straight lines from the chance and effect axes toward the inside of the matrix; the intersection will be the result of the risk as illustrated:

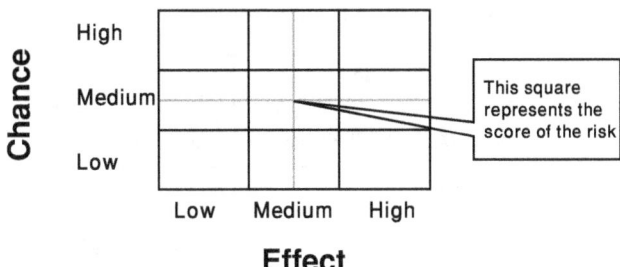

**Effect**

The illustrated risk matrix is the simplest one to use. You an choose from different varieties, such as a bigger matrix with more possible scores for chance and effect or a matrix with numbers instead of quantities (1, 5 and 10 instead of low, medium and high, for example). Also, colored matrices are widely used wherein, for example, red corresponds to a high score, yellow to a medium score and green to a low score.

Now, we need to give meaning (scores) to the internal squares in the risk matrix. This can be done differently, but the most widely used way is illustrated below:

| Risk Score Will Be | When |
|---|---|
| Low | when chance is low and effect is low |
| Medium | when chance is medium and effect is low          or<br>when chance is low and effect is high          or<br>when chance is medium and effect is low          or<br>when chance is medium and effect is medium          or<br>when chance is high and effect is low |
| High | when chance is medium and effect is high          or<br>when chance is high and effect is medium          or<br>when chance is high and effect is high |

Deciding on the scores of risks depends greatly on the importance of the project (risk sensitivity), such that if the project is important and we want to be strict in identifying risks, we will tend to design the matrix to more readily yield higher scores for risks than if the project were of lesser importance.

Many companies have an official risk matrix, so the project manager must look for and use it if such is the case. If no such matrix exists, then the project manager and his/her team should create one, taking into consideration the risk sensitivity of the project.

From the above table, we can rewrite the matrix with the risk scores inside as follows:

| Chance | | Effect | |
|---|---|---|---|
| **High** | Medium | High | High |
| **Medium** | Medium | Medium | High |
| **Low** | Low | Medium | Medium |
| | Low | Medium | High |

**Effect**

- **How to Do Risk Ranking**

**First:** Invite the project team and some experts and possibly stakeholders to a risk ranking meeting.

**Then:** Give everyone a copy of the list of risks and the risk matrix, and explain how it is to be used.

**Next:** Go through the list of risks and, for every identified risk, ask the following two questions:

1 What is the chance that the risk (event) will occur?

2 If the risk does occur, how big of an impact will it have on the project?

**Finally:** Record the score of each risk.

**Example**

The following is the risk ranking for the dinner project using a 3*3 risk matrix:

| Risks | Chance Score | Effect Score | Risk Score |
|---|---|---|---|
| Not finding good quality meat | low | medium | medium |
| Sudden increase in food prices | low | medium | medium |
| The car breaks down | low | medium | medium |
| The meat gets burned during cooking | medium | high | high |
| Not finding good quality potatoes in the market | high | medium | high |
| The gas cylinder goes empty during cooking | low | medium | medium |
| The cafeteria is delayed in delivering the juice | medium | medium | medium |
| The potatoes are undercooked | high | high | high |

Someone might ask: What if, during risk ranking, someone has an idea about how to control a particular risk? Should he/she wait until the next stage to mention it? The answer is No, he/she can mention the idea and the project manager should note it for the next stage. Risk management stages are interrelated, so it is better to record information as soon as it is available.

## Stage Three:  Risk Control

In this stage, we will look for the best ways available to control the identified risks, either by completely removing the risk or by decreasing its magnitude.  The following are four options we can use to tackle risks:

- **Eliminate** the risk completely;

- **Reduce** the magnitude of the risk;

- **Transfer** the risk to another party who can deal with it better than we can;

- **Accept** that the risk might happen.

Before I discuss these options, note the following:

**First:** We need to benefit from the risk ranking activities we did, so rewrite the list of risks starting with the risks that have higher magnitude.  This revised list will help you better use your resources, in that if the risk is high, you will spend more time discussing it and thinking of ways to control it.

**Second:** Remember that risks are in the future, and even if a risk has a very high score, there is still the chance of it not happening.  At this stage, we will start using resources (such as some of the allocated budget), and we want to make sure the data we have concerning risks are of good quality.  To check the quality of your project's risk data, all you have to do is to check whether they were derived in a reasonable and logical manner.  Basically, if you can justify the scoring you have for the risk's chance of occurrence and effect, you are in good shape.  The following is a table for checking the quality of the risk data for the dinner project:

| Risks | Chance | Justification | Effect | Justification |
|---|---|---|---|---|
| Not finding good quality meat | Low | There is no evidence of this in the media | Medium | Alternatives are available |
| Sudden increase in food prices | Low | There is no evidence of this in the media | Medium | Even if there is an increase, it is not expected to be high, and the quantity of food needed is not large |
| The car breaks down | Low | Car was serviced last month | Medium | Can use alternative transportation |
| The meat gets burned during cooking | Medium | Experience in cooking meat is not very good | High | Meat is the key ingredient in the dinner |
| Not finding good quality potatoes in the market | High | Lately, it has been difficult to find good quality potatoes | Medium | Taste will differ |
| The gas cylinder goes empty during cooking | Low | The gas cylinder is new | High | We will not be able to cook |
| The cafeteria is delayed in delivering the juice | Medium | From experience, the cafeteria sometimes is late in its delivery | Medium | We can use soft drinks instead |
| The potatoes are undercooked | High | Experience in cooking potatoes is not very good | High | Potatoes are key ingredients in the dinner |

**Third:** Always keep the risk formula in your mind when thinking of an option to control risks. Consider the example of a 5-year-old girl whose grandmother gave her a bicycle as a gift. The girl's parents are worried the girl will fall and break a leg and want to control this risk. Let's use the risk formula to help the parents ensure the safety of the girl where:

**risk = chance of occurrence * effect**, or

risk of the girl falling = chance of her falling * severity of the fall.

We note from the formula that we can make the risk equal to zero if we make either the chance or severity zero. We can also reduce the magnitude of the risk by decreasing both the chance and severity or one of them. The following table represents decisions parents can make to control the risk:

| Decision | How It Will Reduce the Risk |
|---|---|
| Putting training wheels on the sides of the bicycle | Will reduce risk by reducing the chance of falling |
| Make the child wear protective gear | Will reduce risk by reducing the severity of the fall |
| Putting on training wheels in addition to making the child wear protective gear | Will reduce the risk much more than the two previous decisions |
| Prohibiting the girl from riding the bicycle | This will eliminate the risk entirely |

One moment, please! Deciding on the best option to control the risk is usually a fun exercise, but we also need to consider the "reaction" to each option and balance the positives and negatives. The following table lists some possible reactions to the decisions made for the girl-and-bicycle example:

| Decision | Side Effects |
|---|---|
| Putting on training wheels | -Cost of buying the wheels<br>-Reduction of the joy of riding the bicycle as the wheels will limit the range of motion |
| Making the girl wear protective gear | - Cost of buying the gear<br>- The girl might feel uncomfortable wearing the protective gear<br>- We need someone to make sure the girl actually wears the gear |
| Prohibiting the girl from using the bicycle | - The girl gets angry<br>- Losing the chance of spending good family time<br>- Losing the opportunity for the girl to practice sport |

## Options for Controlling Risks

1 **Risk Elimination:** This might appear to be the best option because it entirely removes the risk by either making its occurrence impossible or such that it will have a negligible effect. For example, there is a risk of undercooking the potatoes in the dinner project example, which can be eliminated by serving steamed vegetables instead. But, by eliminating the potatoes, we will lose the benefit (which is the taste) of serving it. Eliminating risks is not easy to do because it usually costs a lot of money or leads to a loss of some benefits.

2 **Risk Reduction:** You can reduce the magnitude of a risk by either reducing the likelihood of its occurrence or its severity or both. For example, you can reduce the risk of equipment failure by servicing it before you use it. You can also reduce it by providing another piece of equipment on a standby basis so that if the first one fails, the severity will not be very high because you have a replacement as a backup option.

3 **Risk Transfer:** In this case, you arrange for another company or person control the risk for you in return for a fee. You select this option when the other party is more experienced in handling the risk. For example, in the dinner project, you can hire a chef to cook the meat for you so you don't need to worry about burning it. Another general example is hiring a contractor to perform some tasks of the project.

4 **Accepting the Risk:** In many cases you might not be able to use any of the mentioned risk control options due to lack of sufficient funds or lack of specialized companies to transfer risks to. For example, in a risk of equipment failure, you might not be able to use enough money from the project budget to service the equipment. In such cases, we accept that the risk MIGHT happen. However, we can actively accept the risk by making a provision such as allocating extra money to be used only after the risk happens (if it happens at all). For example, we might accept the risk of a possible equipment failure and set aside some amount of the project budget for the rental of a replacement.

To better illustrate the different options, consider the following example of how to use different options for the same risk:

**General Example**

Consider the risk of equipment delay; let us see how you can select from different options to control it. Your strategy selection will depend on factors like the importance of the equipment, the availability of different suppliers and the risk sensitivity of the project.

| Option | Your Goal | How to Implement the Option |
|--------|-----------|------------------------------|
| Avoid | Equipment MUST arrive | If possible, do the work with other equipment that is available (substitute). Alternatively, if possible, eliminate the activity that requires the use of the equipment |
| Reduce | Ensure, as much as possible, that the equipment arrives on time | Reduce the chance of delay by establishing close contact with the supplier. You may also look for other suppliers who, you believe, are more reliable |

| Transfer | Make it the problem of another party | If possible, arrange for a third party to handle the work that requires the use of the equipment |
|---|---|---|
| Accept | You can do nothing to prevent the delay | Either wait for delivery or assign additional money to rent a replacement |

> During the risk control stage, we can make some important decisions, such as changing our project plan and utilizing some money out of our budget (to pay for risk transfer or reduction). For such, it is important to be sure of the quality of the risk data you base your decisions on. Also, you must always weigh the advantages and disadvantages of responding to potential risks or not.

**Opportunities (Good Risks)**

As your project is exposed to many risks, it can also gain many opportunities. For example, in the dinner project, you might have an opportunity to buy meat at a reduced price if you went shopping in the early morning between 7-9 a.m. since the supermarket offers discounts on meat during that time. To capture the opportunity of reduced meat prices, you should:

• **First:** know that there are possible sales, which requires you to purposely think of opportunities;

• **Second:** take the necessary measures to capture the opportunity (by waking up early).

Opportunities can be represented in a similar formula to risks as:

**opportunity = chance of capturing the opportunity * effect of the opportunity**

From the formula, we can expect that the options for harnessing an opportunity are similar to (but opposite of) those used to avoid risks. So we can enhance our chance of capturing an opportunity by increasing either the chance of its occurrence or its effect or both. We might also share opportunities that we could not fully benefit from if we tackled them alone. The following table shows examples of some opportunity-harnessing cases:

| Example | Type of Option |
|---|---|
| In a project to drill an oil rig, the executing company made an agreement with a gas company to capture and use any accompanying natural gas | Sharing |
| In the case that one task is completed early, it will be possible to use the resources to expedite the completion of the next task. As such, the project manager may ask workers to work extra hours | Enhancing the opportunity |

Note that, similar to risk options, opportunity-capturing options also come with side effects. In the first example mentioned in the table, the company will lose some of the profits to the gas company, and in the second example, there will be a need to pay workers for extra hours. However, the gains might outweigh the losses.

> Usually the project team focuses only on thinking about risks to their project with little thought of possible opportunities. For that reason, I have made a special template just to log the opportunities, hoping it will urge the project team to consider them more.

> Because trying to harness opportunities might require some resources, we should select the best opportunities to take advantage of. To do that, we need to rank opportunities, which is a process very similar to the one we used for risks; you can also use a similar matrix for opportunities.

**How to Make a Risk Control Plan**

We go back to brainstorming, where the project manager invites the project team, some experts and some stakeholders to a meeting to discuss possible risk control options. Then the risk options should be logged in a risk register.

**The Risk Register**

The risk register is a piece of paper(s) that "collects" all information about risks in your project.

**Examples**

The following are examples of two registers, one for risks and one for opportunities for the dinner example:

| Risk | Risk Score | Option to Control the Risk | Effect of the Control on the Project | Person Responsible for Implementing |
|------|-----------|---------------------------|-------------------------------------|-------------------------------------|
| Not finding good quality meat | Medium | Accept | - | Project manager |
| Sudden increase in food prices | Medium | Accept | - | Project manager |
| The car breaks down | Medium | Reserve some money to use a taxi | Increase in the budget | Project manager |
| The meat gets burned during cooking | High | Add money to order food from the restaurant | Increase in the budget | Project manager |
| Not finding good quality potatoes in the market | High | Buy ready-to-cook frozen french fries | Increase in the budget<br><br>Delete the task of cutting potatoes | Project manager |
| The gas cylinder goes empty during cooking | Medium | Can use neighbor's cylinder | - | Project manager |
| The cafeteria is delayed in delivering the juice | Medium | Making follow-up calls to the cafeteria | Assign someone for the calls | Wife |
| The potatoes are undercooked | High | Accept | - | Project manager |

The following is the opportunity register for the dinner project:

| Opportunity | Opportunity | | | Option of Capturing Opportunity | Effect of the Option on the Project | Person Responsible for Implementing |
|-------------|-------------|--------|--------------------|---------------------------------|-------------------------------------|-------------------------------------|
| | Chance | Effect | Opportunity Score | | | |
| Buying meat at a reduced price | High | Low | Medium | Wake up early | Minor saving in budget | Project manager |

**Effect of Risk Management on the Project Budget**

After conducting risk management, the project budget is very likely to change. We can represent the new budget by the following formula:

**project budget = revised project original budget + cost of risks that would be controlled + cost of risks that would not be controlled + cost of risks that were not identified (new risks).**

I know the formula seems long, but it is rather straightforward. The following is the explanation of its different elements:

- Revised original project budget: You might change the cost allocation you made for some tasks because you discover errors in estimation. For example, you might discover that the hourly rate of a technician you need for your project is wrong and, thus, change it;

- Cost of risks that can be controlled: This is the money you need in order to transfer or reduce risks (by hiring specialized companies, renting equipment, etc.);

- Cost of risks that could not be controlled: This is the extra money you put aside if you accept a risk (as was discussed in risk acceptance above);

- Cost of risks that were not identified (new risks): Unless you can see into the future, you will not be able to identify all possible risks. You might also identify new risks during project execution; these will require some funding. For those reasons, many companies opt to put aside some extra money in case an unidentified risk emerges. Usually this amount is put in terms of a percentage of the overall project budget, such as 3 or 4%.

If you base your budget on this formula and add up the monetary figures of all its elements, then the accuracy of your estimation will be improved.

> Similarly, the time plan can change after risk management. For example, we might increase the time allotted for a task if we anticipate a risk that the person assigned to the task will be busy with other work at that time.

# Stage Four: Risk Monitoring

Risk monitoring will be done during project execution and will involve the following elements:

- Reviewing the risk register: The risk register should be reviewed to identify which risks have actually happened and how effective the risk control options have been. Doing this will help with capturing lessons learned and will help release resources tied to risks that did not happen. For example, extra money might have been allocated for the risk of an equipment failure; if the equipment did not fail, then this money can be utilized for other purposes;

- Identifying new risks: As the project unfolds, new information will be available and new risks might be discovered. These risks should be identified, ranked, controlled and added to the risk register. New risks can be identified in specific risk meetings or during the project progress report meetings.

**REEM Park Example: Risk Management**

Amanda and her team discussed the park project and decided that the risk sensitivity should be medium. After that, they identified the risks through brainstorming and templates review. A 3*3 risk matrix was used, and they filled out the following template:

Template 17

*Risk Identification and Ranking*

| No. | Risk Event | Chance | Effect | Risk Score |
|-----|-----------|--------|--------|-----------|
| 1 | Insufficient proposals for the design and construction of the park | Low | High | Low |
| 2 | The contracts department gets delayed in reviewing the tender documents | Medium | Medium | Medium |
| 3 | The design of the park is not suitable | Low | High | Medium |
| 4 | Sudden increase in amount of construction materials required | Medium | Low | Medium |
| 5 | Major changes to the scope by the GM | Medium | High | High |
| 6 | The construction contractor goes bankrupt | Medium | Medium | Medium |
| 7 | Underlying service pipes and cables might be present in the park location | Medium | Medium | Medium |
| 8 | Poor communication between the consulting and contracting companies | Medium | High | High |

Template 18

## *Opportunity Identification and Ranking*

| No | Opportunity Event | Chance | Effect | Opportunity Score |
|----|-------------------|--------|--------|-------------------|
| 1 | The tender document for the park construction can be started before the final design is completed (i.e., the two can be done simultaneously, after the completion of the initial design) | Medium | Medium | Medium |
| 2 | Designs of the park can be circulated via email instead of hard copies to expedite reviews | Medium | Medium | Medium |

Template 19

## *The Risk Register*

| No | Risk | Risk Score | Control Option | Responsibility | Side Effect Of The Option On The Project |
|----|------|-----------|----------------|----------------|------------------------------------------|
| 1 | Insufficient proposals for the design and construction of the park | Low | Place an advertisement in newspapers + place ads on the municipality website | Amanda + contracts department | - |
| 2 | The contracts department gets delayed in reviewing the tender documents | Medium | Make sure the documents arrive at the contracts division and follow up on it | Amanda + contracts department | - |

| 3 | The design of the park is not suitable | Medium | Review the initial design, to be done by as many engineers as possible from the municipality | Project Team | There might be slight delays in the review process if too many people are involved |
|---|---|---|---|---|---|
| 4 | Sudden increase in amount of construction materials required | Medium | Design the contract so that the contractor is responsible for any such increase | Contracts department | This might increase the cost of contracting |
| 5 | Major changes to the scope by the GM | High | Actively provide the sponsor (GM) with information about the project and secure his approval on the scope | Amanda | - |
| 6 | The construction contractor goes bankrupt | Medium | Review the financial situation of the company | Project team | - |
| 7 | Underlying service pipes and cables might be present in the park location | Medium | The consultant and contractor must provide a technical risk register for the project to be reviewed by the project team | | |
| 8 | Poor communication between the consulting and contracting companies | High | Conduct quality audits on communication | Amanda | - |

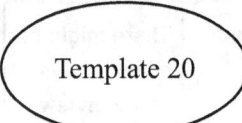

Template 20

*Opportunity Register*

| No | Opportunity | Opportunity Score | Enhancement Option | Responsibility | Side Effects of the Option on the Project |
|----|-------------|-------------------|--------------------|----------------|-----------------------------------------|
| 1 | The tender document for the park construction can be started before the final design is completed (that is, the two can be done simultaneously, after the completion of the initial design) | Medium | Start with the tender draft as soon as the initial design is completed | Amanda | - |
| 2 | Designs of the park can be circulated via email instead of hard copies to expedite reviews | Medium | Obtain soft copies of the designs and make sure that they are sent in a format available to all receivers (like PDF) | Amanda | - |

**Benefits of Using These Templates**

1 You reviewed all previous templates.

2 You identified risks and opportunities, and options to control and capture them.

# Chapter Six
# Project Execution

A main goal of the project book is to provide a plan for your project through templates and to actually use them. Now, as you start executing your project, we need to see how templates can actually be put into use.

**Using the Templates**

There was a lot of intellectual effort put into doing the templates, and we want to transfer it to the project. We can divide the templates into primary and secondary templates as follows:

**Secondary Templates:** These templates were used to fill out other templates but will not be used directly during project execution (for example, the feasibility study and the project map).

**Primary Templates:** These templates will be used directly in execution and include:

* the information sharing plan;
* the time plan;
* the budget;
* the quality plan;
* the risk and opportunity registers;
* the contractors evaluation template;
* the change request form; and
* the project progress form.

# Project Execution Board

The project execution board is a tool to help you collect the information from as many templates as possible into one chart. The idea is simple: Just draw template number 12 ( the time plan) on a big board, then fill it out with information from the other templates as follows:

| Template | Information to Be Used on the Project Board |
|---|---|
| Information sharing plan | From it write the dates of metings and the information to be sent. |
| Budget | From it you can write the required resources in order to prepare them beforehand. |
| Quality plan | You can designate specific dates for quality audits. Also you can write the required quality criteria for every task. |
| Risk and opportunity registers | Enter the expected risks and opportunities for every task. |

**The project board is beneficial because:**

1  It uses visual illustration to show all the project tasks and their attributes;

2  It links all the information from the different templates;

3  It makes it easy to follow the project progress based on time progress;

4  It reduces the number of errors as all the project plan elements are collected in one place and visible to anyone interested in the project.

- **Guidelines for making a project board:**

1  Look for a suitable place in terms of size and accessibility (for example, the wall you choose must not be obstructed by furniture).

2  Try to make the board as big as possible so that the time unit can be small (that is, if you can make the board very big, you can use days or weeks as time units instead of months, which makes tracking task progress easier).

3  Constructing a project board can serve as a team-building exercise, so invite all the team members to participate.

4  Use different colors to write on the board, such as one for risks and one for quality requirements; this should make it easy to navigate through information.

5  You can use sticky notes to write information if you run out of space.

6  To help in project progress tracking, you can use two colored ribbons, one to show today's date and one to show the planned completion date for the task the project is currently on.

**Example: The Project Board for REEM Park**

The following picture is a sample project board for the park project, where the time plan template is enlarged and posted on a wall in Amanda's office.

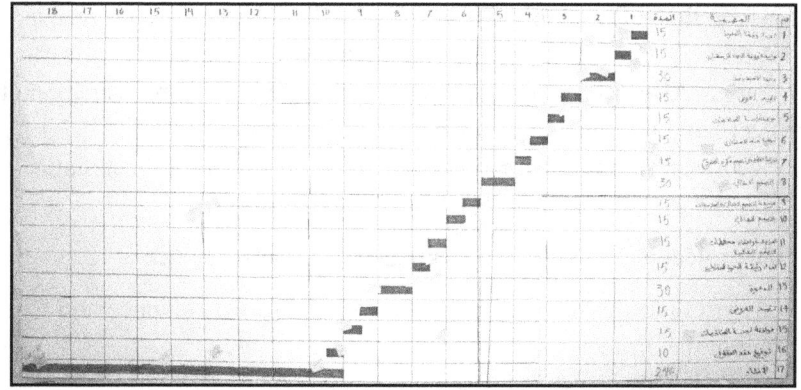

**Note**: The above board is in Arabic so tasks are placed on the right side

### The Project Kick-Off Meeting

How do you declare that your project has entered the execution stage? A good way is to call for a meeting and invite relevant stakeholders. This kick-off meeting will also give you the opportunity to share your plan with intended stakeholders and provide them with essential information, such as informing them that they will have to use a special form for change requests and reminding them of the planned progress meetings.

# How to Use the Templates

### Information Sharing Template

This template will tell you what information you need to send, to whom, when and how. This information will help you prepare to share data as many experts agree that lack of communication is a key reason for project failure. In addition, you will find here information about required meetings and who should attend so you can prepare for them and avoid minor nuisances like the unavailability of meeting rooms. You can also add a record to this template to log the recipients of information in case someone claims he/she did not receive it.

### Time Plan Template

This will be the template you consult on a daily basis. It will help you know what task you need to complete now and what the next step should be. The template will enable you to see the whole project and help you make better decisions when required, such as moving some tasks around in order to expedite the project or compensate for delays. Also, the time plan will be used to track the progress of the project, as will be discussed hereafter:

* **Project Status Reporting**

One of the duties of a project manager is to monitor how the project is progressing in relation to the plan and to share this information with relevant stakeholders. The project sponsor usually makes a big investment in his/her project and needs to be kept informed as to how his/her investment is doing. Also, the need to publicize status reports to relevant stakeholders will prompt the project team to make an effort not to allow its project to slip behind schedule, as it might be viewed as their mistake if it does. To report on the project status, we need the following:

1   A plan for the project to compare to the actual project performance;

2   A calculation system to find the difference between what was planned and what is actually happening within the project;

3   Honesty in providing and analyzing information.

From the list above, you notice that we need to decide on a calculation system. Unfortunately, calculating the difference between planned and actual work is no easy task. It requires a lot of administrative work in documenting and monitoring completed tasks. For example, to find the actual performance of a project in regard to the completion of tasks, we need to know the actual amount of resources used from the total resources and divide the result by the time that has passed. Another thing that adds to the difficulty of calculation is the difference in importance between tasks, regardless of their duration and cost. This is not to say that such calculation is impossible or cannot be used in projects. On the contrary, many projects (especially medium and large ones) adopt a method of calculation known as Earned Value Management. Many publications and case studies exist concerning its use and effectiveness. You may want to search the internet for "Earned Value" and check whether you have the capabilities to use this method in your project.

However, I will present a very simple method that can be used for small to medium sized projects to monitor performance. This method calculates performance based on time by finding the difference between the day the calculation is performed and the date that corresponds to the task the project is working on (as per the time plan). To use this method of calculation, do the following:

1  Use the project board and the time plan.

2  Note today's date.

3  On the project board, decide which task the project is currently on and draw a vertical line upward to determine the date that was planned for this task.

4  Find the difference between the two dates (if the difference is positive, then the project is ahead of schedule; if the difference is zero, then the project is on schedule; and if the difference is negative, then the project is behind schedule).

---

**Question:** Calculate the REEM Park project's progress if today's date is 1 November 2010, and the project is still on the task of evaluating the consultants' proposals.

**Answer:** Using the time plan, we can determine that the task the project is currently on (that is, evaluating consultants' proposals) was planned for 1-15 September 2010. Since the team has not finished the evaluation (as per the question), we can use a date like 7 September. Now we subtract today's date (as given in the question) from the date that corresponds to the task the project is on. So, the progress equals 1 November 2010 minus 7 September 2010 which gives 53 days of delay (approximate).

---

**Note:** Remember that you can use two ribbons on the project board to show the two dates discussed above.

> You can use color-coding to represent the project status. The following is an example:
> Green = if the project is ahead of schedule or there is delay of less than 10 days
> Yellow = if there is delay of 11-29 days
> Red = if there is delay of more than 30 days

## The Budget Template

You can use this template to find the resources needed for every task and, thus, work ahead of time to make sure they are available.

## The Quality Template

This template will enable you to know quality requirements for different tasks, allowing you to plan quality audits for them.

- **Example of Quality Audit:** The REEM Park project team can make a visit to the designing company to make sure required specifications are being incorporated into the design as it progresses.

## The Risk Register

The risk register contains the list of risks and what we need to do to control them. Most of the risk control measures were already incorporated in the project plan when we did the risk control exercise during the planning phase. You can write down the risks and their control measures on the project board to be ready to handle them.

- **Risk Monitoring**

Remember that risks may or may not happen. Regardless of this, we make plans to control them. You need to monitor these risks because if they happen, they might be of a different magnitude than is expected, in which case we will need to make sure the control plans we put in place are still sufficient. Also, recall that we may have decided to accept some risks and reserve some extra money to deal with them should they occur. You need to monitor these risks and the money reserved for them to assess whether it is enough and to release it if the risk does not happen.

- **Identifying New Risks**

As the project unfolds, you might discover possibilities of new risks. These need to be identified, ranked, controlled and logged in the risk register. The money you might need for controlling these new risks can be taken from the money you reserved for unidentified risks when you finalized your project budget during the planning phase. To discover such new risks, you need to hold special meetings for risk identification or put risk identification onto the agenda of the project progress meetings.

**The Opportunity Register**

Similar to the risk register, you need to track opportunities in your project and make sure they are taken advantage of as much as possible to benefit your project.

**Contractors Selection Template**

If the project needs the help of a contractor, we first need to prepare a tender document. This document should describe the required service, its quality requirements and expected duration. You should take your time preparing this document as it will be the basis upon which the contracting company bases its pricing. So try to be as detailed and thorough as possible. To produce a good tender document, you should collect information from the different templates as described in the table below:

| Template to Review | Reason |
|---|---|
| Project Scope | It will be the basis for the tender document. Use it to distinguish what is included in the project from what is not. Think of it as a way to help the contracting company determine the required work and, thus, provide better pricing for your project. |
| Time Plan | To give the contracting company an idea of when you want them to finish (this might also affect the pricing). |
| Quality Plan | Different quality requirements will demand different pricing. |
| Risk Register | Remember that an option in risk control is to transfer the risks. Review the risk register for any risks you want to transfer, and mention them in the tender document. |
| Opportunity Register | Similar to the above point, you might need to share some opportunities. |

Due to the great importance of this document, you should request help from your colleagues or the contracts department if you have little experience in preparing such.

After the document is prepared and sent to potential contractors, the project team will wait for proposals from contractors to deliver the service/s described in the document. To choose the best company, a selection procedure is then started, using the template we made during the planning stage for selecting contractors.

- **REEM Park Example: Selecting Contractors**

Let's assume that Amanda and her team received proposals from three companies for designing the park. Each proposal contained a technical offer of how to design the park and, in a separate envelope, a financial offer detailing the fee the consultant requests. Amanda decided to do the technical evaluation first, so the prices of each company would not affect the technical evaluation. She made copies of the technical offers, gave one to Sara and one to Ali and gave them a week to read them and fill in the evaluation template. The following template illustrates the evaluation done by Amanda, Sara and Ali:

| Comparison Criteria | Weight | Company A | Company B | Company C |
|---|---|---|---|---|
| • Experience of the company in the technical field of the project | 10 | 8 | 9 | 8 |
| • Similar projects conducted | 10 | 9 | 7 | 7 |
| • Experience of the company in the region | 10 | 10 | 7 | 6 |
| • Qualifications of the project team members | 25 | 23 | 20 | 20 |
| • Financial stability | 10 | 10 | 10 | 10 |
| • Compliance with the scope of the service provided by the REEM project team | 25 | 17 | 17 | 16 |
| • Possessing an ISO 9001 certification | 10 | 9 | 9 | 9 |
| Total | 100 | 86 | 79 | 76 |
| Signature | Amanda | | | |

| Comparison Criteria | Weight | Company A | Company B | Company C |
|---|---|---|---|---|
| • Experience of the company in the technical field of the project | 10 | 8 | 9 | 8 |
| • Similar projects conducted | 10 | 9 | 7 | 7 |
| • Experience of the company in the region | 10 | 10 | 8 | 6 |
| • Qualifications of the project team members | 25 | 23 | 21 | 20 |

| | | | | |
|---|---|---|---|---|
| • Financial stability | 10 | 10 | 10 | 10 |
| • Compliance with the scope of service provided by the REEM project team | 25 | 19 | 16 | 16 |
| • Possessing an ISO 9001 certification | 10 | 9 | 9 | 9 |
| Total | 100 | 88 | 80 | 76 |
| Signature | Sara | | | |

| Comparison Criteria | Weight | Company A | Company B | Company C |
|---|---|---|---|---|
| • Experience of the company in the technical field of the project | 10 | 10 | 10 | 9 |
| • Similar projects conducted | 10 | 8 | 9 | 7 |
| • Experience of the company in the region | 10 | 10 | 10 | 6 |
| • Qualifications of the project team members | 25 | 23 | 16 | 17 |
| • Financial stability | 10 | 9 | 9 | 9 |
| • Compliance with the scope of service provided by the REEM project team | 25 | 21 | 22 | 17 |
| • Possessing an ISO 9001 certification | 10 | 9 | 8 | 9 |
| Total | 100 | 90 | 84 | 74 |
| Signature | Ali | | | |

After that, Amanda arranged the scores in another table and computed the final score for each company based on 70% of the total (remember, the technical evaluation weight was determined to be 70% and the financial weight to be 30%).

| Company | Evaluation | | | Average | Final Score (adjusted to be based on 70%) |
|---|---|---|---|---|---|
| | Amanda | Sara | Ali | | |
| A | 86 | 88 | 90 | 88 | 62 |
| B | 79 | 80 | 84 | 81 | 57 |
| C | 76 | 76 | 74 | 75 | 53 |

After that, the financial offers were opened with a representative from the finance department present (to ensure fairness of evaluation). The following table represents costs for each company:

| Company | Price |
|---------|-------|
| A | 1,234,561 |
| B | 1,430,000 |
| C | 1,533,256 |

The criteria for the financial evaluation is cost. The company that will charge the project the lowest cost will get the highest score (in our case, that will be 30% or 30 points); the scores of the other two companies will be determined relative to the lowest scoring company by this formula: financial score = (price of the company / lowest price) * (weight of the financial evaluation). The following table lists the results:

| Company | Price | Financial Evaluation |
|---------|-------|----------------------|
| A | 1,234,561 | 30 points, because they have the lowest bid |
| B | 1,430,000 | (1430000/1234561)*30 = 26 points |
| C | 1,533,256 | (1533256/1234561)*30 = 24 points |

Now we can add the scores from the technical and financial evaluations and find the highest score, which will determine the winning company.

| Company | Technical Evaluation Score (out of 70%) | Financial Evaluation Score (out of 30%) | Final Score (out of 100%) |
|---------|------------------------------------------|------------------------------------------|----------------------------|
| A | 62 | 30 | 92 |
| B | 57 | 26 | 83 |
| C | 53 | 24 | 77 |

Company A wins with a score of 92%.

## Project Progress Reporting

Progress reporting is among the best ways to communicate information while executing your project. It is best to generate these reports in progress meetings where stakeholders are present. These meetings can serve as a stakeholders management tool because they allow for the expression of their concerns and make them feel important to the project team.

- **REEM Park Example: Progress Reporting**

This is an example of what a completed progress report might look like in the case of the REEM Park project:

| Report Number: 14 Date: 1 August 2010 | |
|---|---|
| Project Status | On track |
| Days of Delay | None |
| Tasks Completed This Month | Finished the outside fence Working to link sewage system with the mainframe |
| Risks Facing the Project | - |
| New Risks | Due to the summer heat, the work time will be reduced by 2 hours as per the instruction of the Health and Safety Authority. Communication with the contractor is ongoing to compensate for the lost work time. |
| Upcoming Tasks | Install foundations for the fountain Continue to link the sewage systems |
| Lessons Learned | - |
| Project Manager Signature | Amanda |

## Change Requests

As an example, let's assume that Amanda received a phone call from the GM's office informing her that the GM would like to add shading for the children's playground area. Because the request came from the sponsor (represented by the GM), Amanda gave it her direct attention and called for a meeting with the GM and the manager of the parks department to fill in the change request template and log the reasons behind it. The GM explained his concerns that without the canopies, children using the playground would be facing the danger of heat exhaustion. He also added that the canopies will prolong the life expectancy of the playground equipment.

Then Amanda contacted the contractor and asked for a quote on the canopies and how long they would take to install. The contractor informed her that the canopies were available in the market and could be installed simultaneously with preparing the playground and thus should not affect the project schedule. As for their cost, the contractor said they would cost around $30,000 US.

Amanda checked the project budget and found that the cost of the canopies could not be accommodated from the project reserves. She contacted the manager of finance and agreed to issue an order for this amount as long as it was approved by the GM. The following is an example of the change request for this scenario:

| Change Request Form No. 1<br>REEM Park Project | |
|---|---|
| Change description | Add canopies to the playground in REEM Park |
| Reasons for change | To protect children and equipment from sun heat |
| Name of person submitting | Municipality GM |
| Date | 12 July 2010 |
| Possible effects if the request is accepted | Increase the project cost by $30,000 |
| Is the request approved? | Yes |
| Signature of Project Manager | Amanda |

# Chapter Seven
# Project Closure

# Lessons Learned

Recall that we started planning for our project by reviewing lessons learned from past projects. Now it's our project's turn to add to the knowledge base of the organization. By "lessons learned," I mean the information that we discovered during the project's life cycle which affected it either positively or negatively. By sharing this information, the project team believes it will help future project teams to not make similar mistakes and not miss out on opportunities.

The importance of the lessons learned comes from being very specific to the organization. They should enable us to discover the organization's shortcomings and fix them in a way that cannot be achieved through any other means. For example, such information cannot be learned by reading management books or through general training. Allow me to ask you the following question: If you were to buy a management book on communication and found two books you liked, and the author of the first book works at your organization while the author of the second book works at a different one, which book would you buy? Most likely, you will opt for the one written by the author from your organization since you assume he/she is likely to give specific examples from which you will benefit, more so than the other, who will undoubtedly provide less familiar examples. The same reasoning can be used to show that lessons learned are uniquely valuable to an organization than general knowledge because they arise from day-to-day work and its specific needs. Examples of lessons learned might be the following:

• After trying a contracting company, you discover that they are difficult to deal with;

• Identifying better ways of working in a particular area;

• With new information, you can make better estimates of task duration and cost;

• Recommendations to amend certain templates used by the organization;

• Discovering a particular area in which the organization could benefit from additional training.

## Guidelines for Collecting Lessons Learned

1 Invite the project team and representatives from the stakeholders and contractors to a lessons-learned collection meeting.

2 Start brainstorming with the attendees about possible lessons. In many cases, lessons learned will be the result of mistakes made by certain people; as such, you will want to control the meeting in such a way that the focus is on the mistakes, not the people who made them.

3 Try to categorize lessons learned under general areas, such as communication, project management, internal procedures, contractors, etc.

4   Start thinking with the attendees about practical ways to actually benefit from the lessons learned. We can separate lessons learned into the following types:

- Lessons learned that can be implemented quickly: These lessons learned usually fix a problem in a narrow area and don't require funding. For example, you might discover that the templates used by the finance department to ask for extra funding require some amendments. As a result, you can send your recommendations to the finance department through an official letter and, thus, you have "satisfied" the lesson learned.

- Lessons learned that require approval and funding: An example might be a recommendation arising from a project that a project management office be created whose responsibility is to provide support to project managers. Such a recommendation will usually take a while to study and, if approved, it will become a project in itself.

5   Log the lessons in the Lessons Learned Template and distribute them to relevant sections in your company to follow on their implementation (like the Knowledge Management Office or the PMO).

**REEM Park Example: Lessons Learned**

The following is an example of the Lessons Learned Template for the park project.

Template 21

*Lessons Learned*

| No | Lesson Learned | Category | How to Share/Implement the Lesson |
|---|---|---|---|
| 1 | The design submitted by the consulting firm was very good | Contracts | Put the name of the company on the preferred contractors list |
| 2 | The contracting company did not provide a proper risk register since it wasn't mentioned as a requirement in the contract document | Contracts | To specify in future contracts that a risk register will be required |

| 3 | During summertime, the workday is reduced by 2 hours | Legal | Send a copy of the regulation to all project managers in the organization for them to rethink their time plans if their projects will take place in the summertime |
|---|---|---|---|
| 4 | Using artificial grass in the park helped reduce the overall construction period | Technical - Parks | Share the information within the parks department |
| 5 | Many of the water sprinklers used were defective out-of-the-box | Technical - Parks | Avoid these types of sprinklers in the future.  Share the information within the parks department |

Quite often, the lessons-learned process stops at filling out the templates without any real follow-up on their implementation.  To solve this, many companies have created what is known as a project management office, one of whose important tasks is to archive lessons learned, categorize them and follow up on their implementation. Also, sometimes this task is given to the knowledge management office.

It will be a good idea to invite decision-makers to your lessons-learned meeting, since they might be able to implement some of the lessons on-the-spot.

Collecting lessons learned is an ongoing process done throughout the life cycle of the project.  So make sure you add lessons to your template as they arise.

You also have the option of collecting lessons learned "virtually" through email.

**Benefits of Using This Template**

1  You participated in the process of knowledge capturing and sharing in your organization.

2  You made suggestions (built on actual experience) to improve future projects.

3  You gave stakeholders the chance to participate in the project (by giving their opinions about what went well and what did not).

# Project Team Assessment

As a project manager, you can provide valuable comments to your team on their performance. Some of the members of your project team might have come from other departments, in which case your assessment might play a big part in their yearly evaluation. To make a fair assessment, you need to have a benchmark for comparing members' performance. Your basis will be the Project Team Plan, where you listed the responsibilities of each member and had them sign it.

**REEM Park Project: Team Member Evaluation**

The following is an example of a member evaluation form completed by Amanda for one of the team members:

Template 22

*Project Team Member Evaluation Form*

| REEM Park Project | |
|---|---|
| Name of Team Member | Sara |
| Main Tasks | Help the project manager in preparing the Project Book binder |
| Did the Member Complete All Tasks? | Yes |
| Communication within the Team | Very Good |
| Communication with Stakeholders | Very Good |
| Level of Organization | Very Good |
| Knowledge of Project Management Skills | Good |
| Comments | Sara did an excellent job in planning for the project. In addition, she attended all the project progress meetings and helped a lot in collecting the lessons learned. |
| Project Manager | Amanda |
| Date | 19-12-2011 |

## Benefits of Using This Template

You identified and acknowledged good (or poor) performance within your project team.

# Project Closure Template

Let's go back to the definition of a project as a collection of tasks with a start and an end. Now is the time to end our project. You will be amazed to know how many projects are finished without any official document that describes that the project is officially concluded and that resources have been released. Our goal in filling out this template is to secure project sponsor approval of what the project has produced and to release people and equipment from the project.

> You can fill out the closure template during a celebration to announce that the project was successful and to distribute some certificates of achievement to distinguished team members and contractors.

## REEM Park Project: Closure Template

This is an example of a project closure template for the park signed by the manager of the parks department on behalf of the general manager.

Template 23

*Project Closure Template*

| Project Duration | | Budget | |
|---|---|---|---|
| Planned | Actual | Planned | Actual |
| 510 days | 520 days | $2 million US | $1.7 million US |
| Was the project completed as per the original scope of work? | | | Yes |
| Have all resources been released? | | | Yes |
| Signature of the Sponsor | | | M.P |
| Signature of the Project Manager | | | Amanda |

**Benefits of Using This Template**

1  You secured official acceptance from the sponsor on the project products.

2  You officially closed the project and, thus, can release resources.

# Final Word

My goal in this book has been to present project management in a simple way and encourage project managers to adopt the various methods and tools presented herein.  The Project Book is a framework that takes you step by step toward successful project completion by providing a template for each of the elements in a project.  Filling out the templates relies heavily on brainstorming with other people, thus providing new and unique ideas for managing the project at hand.  I have also focused on risk management as a means of pulling everything together to ensure harmony among the different templates, thus producing a more realistic and trustworthy project plan.

As I stated at the outset, Project Management is a separate practice from the technical field of the project.  I advise you to explore it further as it will help you manage and achieve successful projects.  I sincerely hope you will benefit from this book.

To offer comments, please contact me at this email: **alk.books@gmail.com.**  Remember, you can download the templates contained throughout this book from the following website: **www.theprojectbook.net or www.kuwaitat.net**

# Chapter Eight
# Exercises

In this chapter you will have a chance to practice producing a Project Book. There are two examples:

- **Example 1:** to make a Project Book for constructing a website

- **Example 2:** to make a Project Book for a family trip to Paris

For each example I will provide the following:

- A simple scenario;

- A set of empty templates to fill out;

- An example of a solution.

**Note 1:** When filling out the templates you will notice an empty page next to each one. You can use it for your brainstorming and notes.

**Note 2:** The solved examples are for the purpose of illustration and comparison only, there is no one way to fill out a template.

# Example One: Building a website

**Scenario**: You work as a public relations officer in a large furniture manufacturing company. Your boss just came from a meeting with the general manager and told you that you have been assigned as a project manager to create a website for the company. The goal of the site is to provide an online catalogue and receive customers' comments.

## Assumptions:

- You will be managing a team of three people, an IT engineer from the IT department and two fresh graduates working as assistants.

- You will need to hire a company to build the site for you.

- You have support from top management to finish the site as soon as possible.

- You may make additional assumptions.

**Now, start filling out the templates!**

# Use This Page for Brainstorming and Notes

Template 1

# Non Financial Return on Investment

| Expected Return | Chance of Attaining the Return | Link with Organizational Goals | Cost |
|---|---|---|---|
| | | | |
| | | | |
| | | | |
| | | | |
| | | | |
| | | | |
| | | | |
| | | | |

# Use This Page for Brainstorming and Notes

Template 2

# Feasibility Study

| Description of the Idea |
|---|
| |
| Available Alternatives |
| |
| Availability of Funding |
| |
| Can the Idea Be Implemented Technically? |
| |
| Availability of Human Resources to Run the Project |
| |
| Conflicts with Other Projects |
| |
| Recommendations |
| |

# Use This Page for Brainstorming and Notes

Template 3

# Project Charter

| | |
|---|---|
| Project Name | |
| Project ID | |
| Expected Duration | |
| Expected Budget | |
| Project Goals | |
| Project Owner | |
| Project Manager | |
| Start Date | |
| Signature | |

# Use This Page for Brainstorming and Notes

Template 4

# Project Team

| No. | Name | Tasks | Signa-ture |
|-----|------|-------|------------|
|     |      |       |            |
|     |      |       |            |
|     |      |       |            |
|     |      |       |            |
|     |      |       |            |
|     |      |       |            |
|     |      |       |            |
|     |      |       |            |
|     |      |       |            |

# Use This Page for Brainstorming and Notes

Template 5

# Lessons Learned Review

| No. | Previous Project Name | Lesson Learned | How to Use It |
|---|---|---|---|
| | | | |
| | | | |
| | | | |
| | | | |
| | | | |
| | | | |
| | | | |
| | | | |
| | | | |

# Use This Page for Brainstorming and Notes

Template 6

## Stakeholders Identification and Prioritization

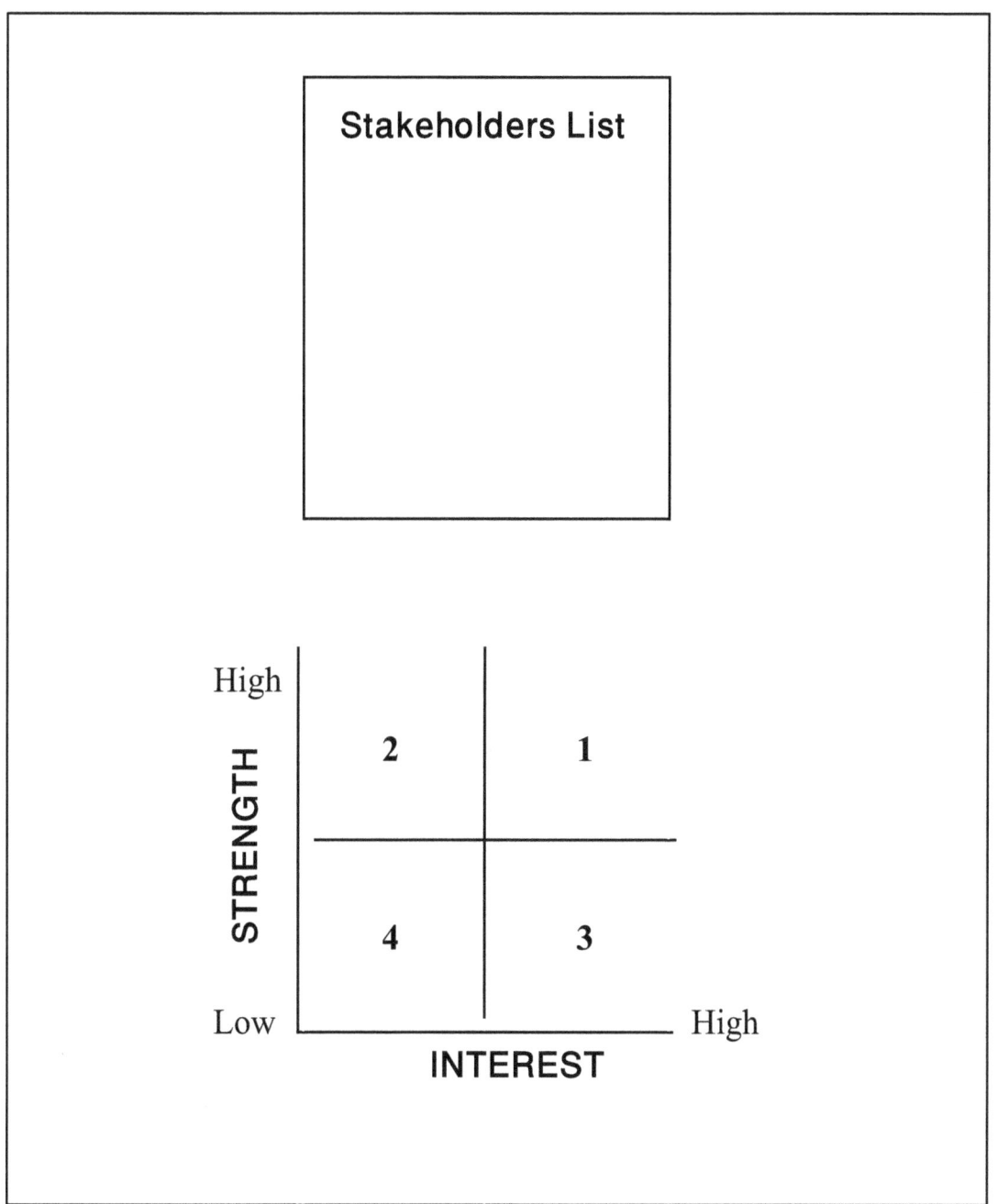

# Use This Page for Brainstorming and Notes

Template 7

## Stakeholders Requirements

| Stakeholders | Potential Requirements | How to Satisfy Requirements |
|---|---|---|
| | | |
| | | |
| | | |
| | | |
| | | |
| | | |
| | | |
| | | |
| | | |

# Use This Page for Brainstorming and Notes

Template 8

# Information Sharing Plan

| Information | Receivers | Method of Sending | Frequency of Sending | Who Will Send the Information? |
|---|---|---|---|---|
| | | | | |
| | | | | |
| | | | | |
| | | | | |
| | | | | |
| | | | | |
| | | | | |
| | | | | |
| | | | | |

| Receiver | Contact | | Receiver | Contact |
|---|---|---|---|---|
| | | | | |
| | | | | |
| | | | | |

# Use This Page for Brainstorming and Notes

Template 9

# Project Scope

Project Picture

What's Out

What's In

# Use This Page for Brainstorming and Notes

Template 10

## Project Map

```
  ╭─────────────────────╮
  │                     │
  │      website        │
  │                     │
  ╰─────────────────────╯
```

**Use This Page for Brainstorming and Notes**

Template 11

# Project Quality Requirements

| No. | Task | Required Quality | How to Measure It |
|---|---|---|---|
|  |  |  |  |
|  |  |  |  |
|  |  |  |  |
|  |  |  |  |
|  |  |  |  |
|  |  |  |  |
|  |  |  |  |
|  |  |  |  |
|  |  |  |  |

# Use This Page for Brainstorming and Notes

Template 12-A

## Project Map with Durations of Tasks

```
┌─────────────────────┐
│                     │
│      website        │
│                     │
└─────────────────────┘
```

## Use This Page for Brainstorming and Notes

Template 12-B

**Network Diagram**

```
┌─────────┐                              ┌─────────┐
│  START  │                              │   END   │
└─────────┘                              └─────────┘
```

## Use This Page for Brainstorming and Notes

Template 12

# Project Time Plan

**Time unit is ....**

| No. | Task | Time | 1 | 2 | 3 | 4 | 5 | 6 | 7 | 8 | 9 | 10 | 11 | 12 | 13 | 14 | 15 | 16 | 17 | 18 |
|-----|------|------|---|---|---|---|---|---|---|---|---|----|----|----|----|----|----|----|----|----|
|     |      |      |   |   |   |   |   |   |   |   |   |    |    |    |    |    |    |    |    |    |
|     |      |      |   |   |   |   |   |   |   |   |   |    |    |    |    |    |    |    |    |    |
|     |      |      |   |   |   |   |   |   |   |   |   |    |    |    |    |    |    |    |    |    |
|     |      |      |   |   |   |   |   |   |   |   |   |    |    |    |    |    |    |    |    |    |
|     |      |      |   |   |   |   |   |   |   |   |   |    |    |    |    |    |    |    |    |    |
|     |      |      |   |   |   |   |   |   |   |   |   |    |    |    |    |    |    |    |    |    |
|     |      |      |   |   |   |   |   |   |   |   |   |    |    |    |    |    |    |    |    |    |
|     |      |      |   |   |   |   |   |   |   |   |   |    |    |    |    |    |    |    |    |    |
|     |      |      |   |   |   |   |   |   |   |   |   |    |    |    |    |    |    |    |    |    |
|     |      |      |   |   |   |   |   |   |   |   |   |    |    |    |    |    |    |    |    |    |
|     |      |      |   |   |   |   |   |   |   |   |   |    |    |    |    |    |    |    |    |    |
|     |      |      |   |   |   |   |   |   |   |   |   |    |    |    |    |    |    |    |    |    |
|     |      |      |   |   |   |   |   |   |   |   |   |    |    |    |    |    |    |    |    |    |
|     |      |      |   |   |   |   |   |   |   |   |   |    |    |    |    |    |    |    |    |    |
|     |      |      |   |   |   |   |   |   |   |   |   |    |    |    |    |    |    |    |    |    |
|     |      |      |   |   |   |   |   |   |   |   |   |    |    |    |    |    |    |    |    |    |

# Use This Page for Brainstorming and Notes

Template 13

## Resources Table

| No. | Task | Human Resources | Cost/Comments | Equipment/ Materials | Cost/ Comments |
|-----|------|-----------------|---------------|---------------------|----------------|
|     |      |                 |               |                     |                |
|     |      |                 |               |                     |                |
|     |      |                 |               |                     |                |
|     |      |                 |               |                     |                |
|     |      |                 |               |                     |                |
|     |      |                 |               |                     |                |
|     |      |                 |               |                     |                |
|     |      |                 |               |                     |                |
|     |      |                 |               |                     |                |
|     |      |                 |               |                     |                |
|     |      |                 |               |                     |                |
| Total |    |                 |               |                     |                |
| Project Budget |  |          |               |                     |                |

# Use This Page for Brainstorming and Notes

Template 14

# Technical Evaluation Template for Web Project

| Comparison Criteria | Weight | Company 1 | Company 2 | Company 3 | Company 4 | Company 5 |
|---|---|---|---|---|---|---|
|  |  |  |  |  |  |  |
|  |  |  |  |  |  |  |
|  |  |  |  |  |  |  |
|  |  |  |  |  |  |  |
|  |  |  |  |  |  |  |
|  |  |  |  |  |  |  |
|  |  |  |  |  |  |  |
|  |  |  |  |  |  |  |
| Total |  |  |  |  |  |  |
| Signature |  |  |  |  |  |  |

Template 15

# Make Your Own Change Request Template

Template 16

# Make Your Own Progress Reporting Template

## Use This Page for Brainstorming and Notes

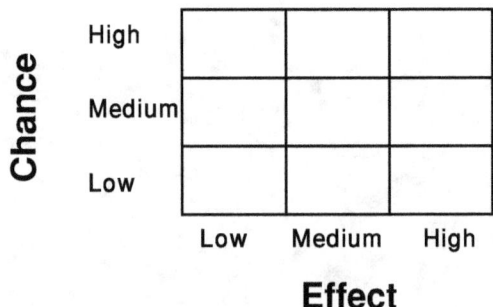

Template 17

# Risk Identification and Ranking

| No | **Risk** Event | Chance | Effect | Risk Score |
|----|----------------|--------|--------|------------|
|    |                |        |        |            |
|    |                |        |        |            |
|    |                |        |        |            |
|    |                |        |        |            |
|    |                |        |        |            |
|    |                |        |        |            |
|    |                |        |        |            |
|    |                |        |        |            |
|    |                |        |        |            |
|    |                |        |        |            |

# Use This Page for Brainstorming and Notes

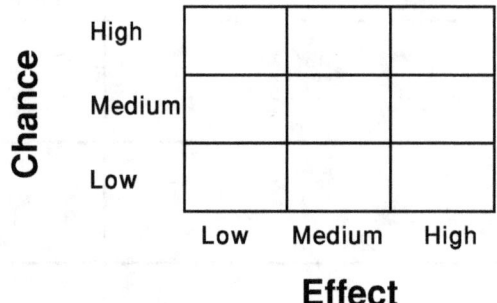

Template 18

# Opportunity Identification and Ranking

| No | Opportunity Event | Chance | Effect | Opportunity Score |
|----|-------------------|--------|--------|-------------------|
|    |                   |        |        |                   |
|    |                   |        |        |                   |
|    |                   |        |        |                   |
|    |                   |        |        |                   |
|    |                   |        |        |                   |
|    |                   |        |        |                   |
|    |                   |        |        |                   |
|    |                   |        |        |                   |
|    |                   |        |        |                   |

# Use This Page for Brainstorming and Notes

Template 19

# The Risk Register

| No. | Risk | Risk Score | Control Option | Responsibility | Side Effect of the Option on the Project |
|-----|------|------------|----------------|----------------|------------------------------------------|
|     |      |            |                |                |                                          |
|     |      |            |                |                |                                          |
|     |      |            |                |                |                                          |
|     |      |            |                |                |                                          |
|     |      |            |                |                |                                          |
|     |      |            |                |                |                                          |
|     |      |            |                |                |                                          |
|     |      |            |                |                |                                          |
|     |      |            |                |                |                                          |

# Use This Page for Brainstorming and Notes

Template 20

# Opportunity Register

| No | Opportunity | Opportunity Score | Enhancement Option | Responsibility | Side Effects of the Option on the Project |
|----|-------------|-------------------|--------------------|----------------|-------------------------------------------|
|    |             |                   |                    |                |                                           |
|    |             |                   |                    |                |                                           |
|    |             |                   |                    |                |                                           |
|    |             |                   |                    |                |                                           |
|    |             |                   |                    |                |                                           |
|    |             |                   |                    |                |                                           |
|    |             |                   |                    |                |                                           |
|    |             |                   |                    |                |                                           |

# A Sample Project Book for Building a website

Template 1

## Non-Financial Return on Investment

| Expected Return | Chance of Attaining the Return | Link with Organizational Goals | Cost |
|---|---|---|---|
| Serve as a marketing tool for the company and its products | High | High | Money needed for the company building the site |
| The possibility of future e-business (for example, online store) | Medium | Medium | Time dedicated to the project by the project team |
| If customers' comments are received online, they can be responded to faster | High | High | Money for buying a web domain and storage space |
| Can use the site for additional services (for example, recruitment) | Medium | Medium | Assigning a web administrator to maintain the website |

Template 2

## Feasibility Study

| Description of the Idea |
|---|
| Build a website to publish the company product catalogue |
| Available Alternatives |

| Distributing the product catalogue through mail, but it will be more expensive |
| --- |
| Availability of Funding |
| A rough estimation of the required budget is around $5000 US. Funding is available through the Public Relations and IT departments |
| Can the Idea Be Implemented Technically? |
| Yes |
| Availability of Human Resources to Run the Project |
| Project manager will be provided from the Public Relations Department. |
| Conflicts with Other Projects |
| No conflicts are present. The IT Department will be involved in the project and the general manager is fully supporting the idea. |
| Recommendations |
| Go ahead and start planning for the project |

Template 3

## Project Charter

| Project Name | Company website |
| --- | --- |
| Project ID | 23-PRD-10 |
| Expected Duration | 45 days |
| Expected Budget | $5000 US |
| Project Goals | Build a website that provides information about the company and its product catalogue |
| Project Owner | Public Relations Department |
| Project Manager | Bill |
| Start Date | 15-3-2010 |
| Signature | The GM |

Template 4

## Project Team

| No. | Name | Tasks | Signature |
|---|---|---|---|
| 1 | Bill | Project Manager. Responsible for completing the Project Book binder and all aspects of the project during the different stages. Also directly responsible for managing the web designing company and responsible for deployment of the site | *bill* |
| 2 | Tony | Help in completing the Project Book binder. Design a draft of the web layout. Help with testing. Provide training for the website administrator | *tony* |
| 3 | Sally | Help in completing the Project Book binder. Collect the required information about the company and its products | *sally* |
| 4 | Kurash | Help in administrative work, such as preparing documents and memos and calling for meetings. Help in completing the Project Book binder. Collect the required information about the company and its product | *kurash* |

Template 5

## Lessons Learned Review

| No. | Previous Project Name | Lesson Learned | How to Use It |
|---|---|---|---|
| 1 | No previous lessons learned found | | |

Template 6

## Stakeholders Identification and Prioritization

| Stakeholders List |
| --- |
| 1 GM |
| 2 Public Relations department |
| 3 IT department |
| 4 Customers |

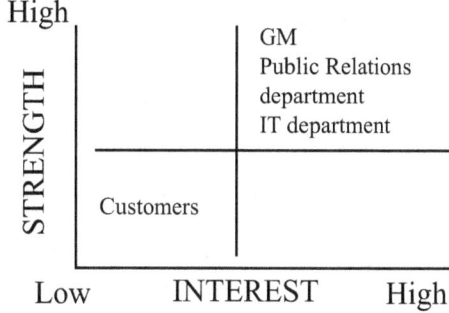

Template 7

## Stakeholders Requirement

| Stakeholders | Potential Requirements | How to Satisfy Requirements |
| --- | --- | --- |
| GM | Project is completed within time and budget. The design to be top-notch. | By proper planning. Also, the GM must approve the web design and be kept updated on project progress |
| Public Relations Department | The project to be finished on time and the GM likes the outcome. Also, all relevant information about the company to be included (for example, its history, services and products). | Through proper planning. Also, to make sure all relevant data are gathered and the site is tested before it goes online. |
| IT Department | To be involved in selecting the designing company. Also, to be part of writing the technical requirements for the site and final approvals. | This will be accommodated by having a representative of the IT department on the project team. |
| Customers | For the site to be easy to navigate through. Also, to have information about the company and its products and to be able to send comments online | These requirement to be put in the tender document and the selected designing company to make sure it takes them into consideration. |

Template 8

## Information Sharing Plan

| Information | Receivers | Method of Sending | Frequency of Sending | Who Will Send the Information |
|---|---|---|---|---|
| Risk Identification Meeting | project team + representative from IT department | Email | One time | Kurash |
| Time Plan | project team + GM + Public Relations manager + IT manager | Official memo | One time and after any change | Kurash |
| Kick-Off Meeting with the contractor | project team + manager of parks department + project management office | Email | One time | Kurash |
| Preliminary Design of the Site | project team + GM + Public Relations manager + IT manager | Email + mail | One time | Kurash |
| Final Design of the Site | project team + GM + Public Relations manager + IT manager | Official memo | One time | Kurash |
| Project Progress Report | project team + contractor | Email + mail | First week of every month | Kurash |
| Lessons Learned Meeting | project team + contractor | Email | One time | Kurash |

| Receiver | Contact | | Receiver | Contact |
|---|---|---|---|---|
| GM | Secretary | | IT Department | Tony |
| Public Relations Department | Bill | | | |

Template 9

## Project Scope

**Picture of Project Deliverable (Components)**

- Picture of the company workshop on the background
- Button for "About the Company"
- Button for the product catalogue
- Button for "Contact Us"
- Button for "Picture Gallery"
- Button for "Video Gallery"

**What's Out**

- Online store
- Forum and chatting

**What's In**

- Information about the company
- Information about the products
- Pictures and videos
- Ability to send emails

Template 10

## Project Map

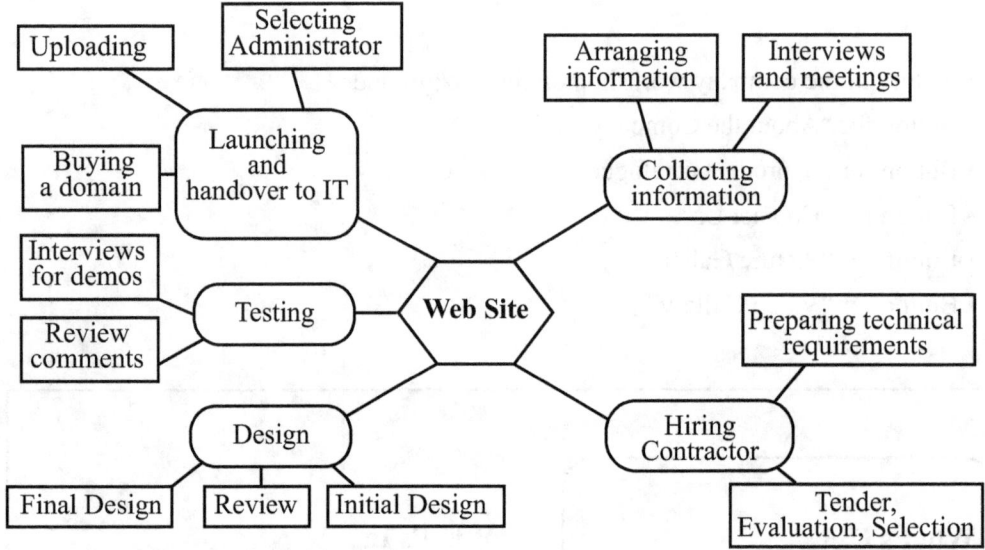

Template 11

## Project Quality Requirements

| No. | Task | Required Quality | How to Measure It |
|---|---|---|---|
| 1 | Collecting Information | Collect all information | Information to be collected from all over the company (a letter can be sent from the public relations department to all other departments requesting information that might be relevant for the website). After information is collected, a review of its completeness should be conducted. |
| 2 | Hiring Contractor | Select the best company | By using a clear method to assess and select among the different companies |
| 3 | Design | To be both nice to view and easy to navigate through | By asking the designing company to conduct usability testing |
| 4 | Testing | Test the site with real customers and review their requirements | Make sure to prepare a log for testing, to include the name of person and his/her comments. Also, in the time plan allow for sufficient time for testing and reviewing comments |
| 5 | Launching and Hand-Over to IT | Combine launching with a mini marketing campaign that includes press releases and printing the web address on company catalogues. Also, a suitable web name to be selected | Prepare some web press releases. Also, print the web name on all upcoming catalogues |

Template 12 -A

## Project Map with Durations for Each Task

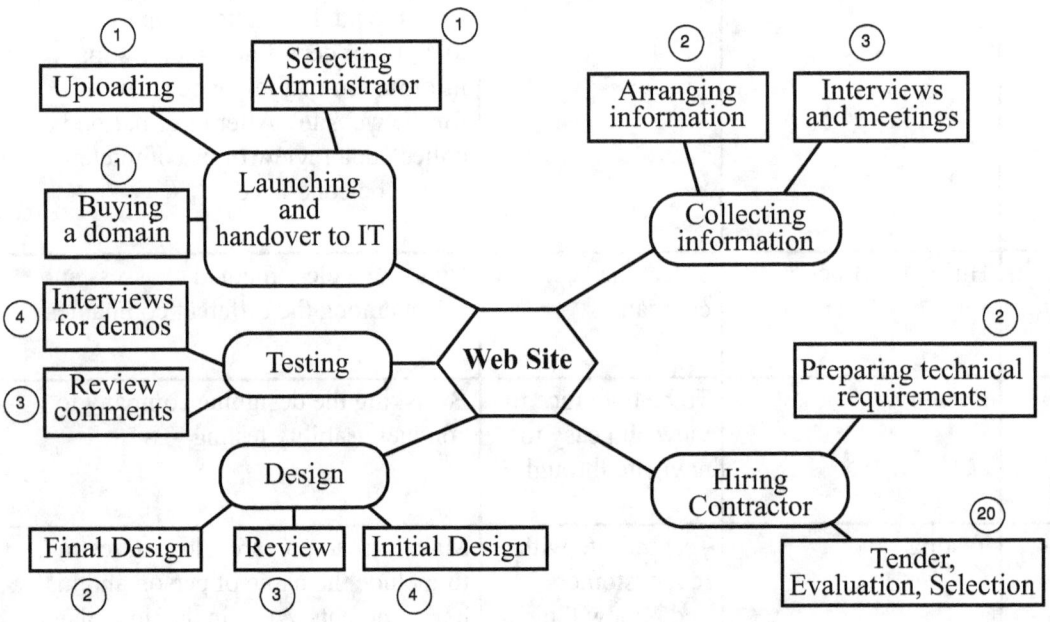

Template 12 -B

## Network Diagram

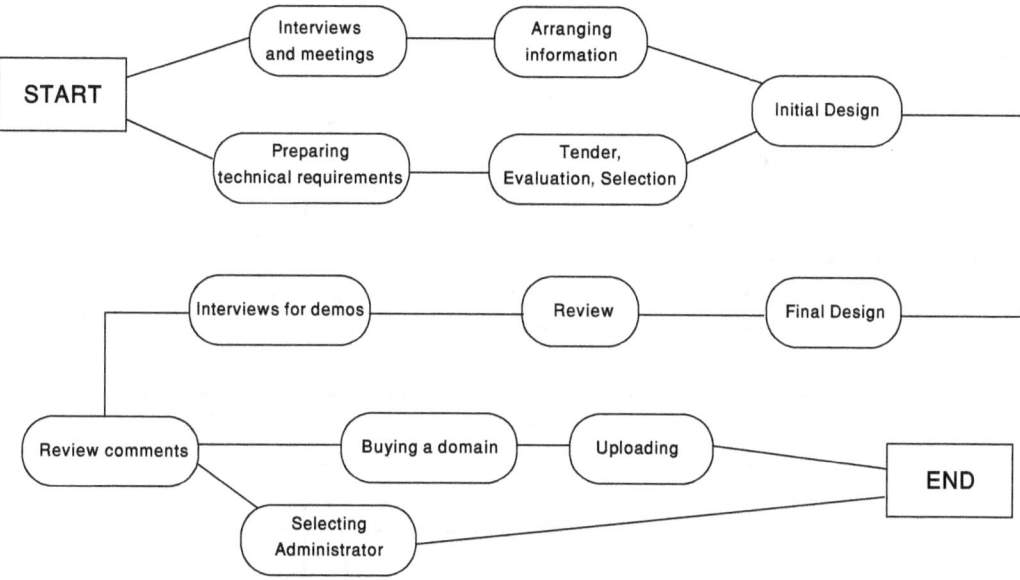

Template 12

## Project Time Plan

Time unit is days

| No. | Task | Dura-tion | 5 | 5 | 5 | 5 | 5 | 5 | 5 | 5 | 5 | 5 | 5 |
|---|---|---|---|---|---|---|---|---|---|---|---|---|---|
| | **Collecting Information** | | | | | | | | | | | | |
| 1 | Interviews and Meetings | 3 | ▪ | | | | | | | | | | |
| 2 | Arranging Information | 2 | ▪ | | | | | | | | | | |
| | **Hiring Contractor** | | | | | | | | | | | | |
| 3 | Preparing Technical Requirements | 2 | | ▪ | | | | | | | | | |
| 4 | Tender, Evaluation, Selection | 20 | | ▬▬▬▬ | | | | | | | | | |
| | **Design** | | | | | | | | | | | | |
| 5 | Initial Design | 4 | | | | | | ▪ | | | | | |
| 6 | Review | 3 | | | | | | | ▪ | | | | |
| 7 | Final Design | 2 | | | | | | | ▪ | | | | |
| | **Testing** | | | | | | | | | | | | |
| 8 | Interviews for Demos | 4 | | | | | | | | ▪ | | | |
| 9 | Review Comments | 3 | | | | | | | | | ▪ | | |
| | **Launching and Hand-Over to IT** | | | | | | | | | | | | |
| 10 | Buying a Domain | 1 | | | | | | | | | | ▪ | |
| 11 | Uploading | 1 | | | | | | | | | | ▪ | |
| 12 | Selecting Administrator | 1 | | | | | | | | | | ▪ | |

Template 13

## Resources Table

| Task | Human Resources | Cost/ Comments | Equipment/ Materials | Cost/ Comments |
|---|---|---|---|---|
| **Collecting Information** | | | | |
| Interviews and Meetings | project team | - | - | - |
| Arranging Information | project team | - | - | - |
| **Hiring Contractor** | | | | |
| Preparing Technical Requirements | project team | - | - | - |
| Tender, Evaluation, Selection | one person | - | - | - |
| **Design** | | | | |
| Initial Design | contractor | $4000 US | - | - |
| Review | project team | - | - | - |
| Final Design | contractor | - | - | - |
| **Testing** | | | | |
| Interviews for Demos | contractor + project team | - | laptop | available |
| Review Comments | contractor + project team | - | - | - |
| **Launching and Hand-Over to IT** | | | | |
| Buying a Domain | project team | $1000 US | - | - |
| Uploading | contractor | - | - | - |
| Selecting Administrator | project team + IT manager | - | - | - |
| **Total** | $5000 US | | | |

Template 14

## Technical Evaluation Template for Web Project

| Comparison Criteria | Weight | Company 1 | Company 2 | Company 3 | Company 4 | Company 5 |
|---|---|---|---|---|---|---|
| Quality of and number of similar websites created in the past | 50 | | | | | |
| Qualifications of the project team that will be provided | 20 | | | | | |
| Compliance with the scope of service provided by the project team | 25 | | | | | |
| Possession of an ISO 9001 certification | 5 | | | | | |
| Total | 100 | | | | | |
| Signature | | | | | | |

Template 17

## Risk Identification and Ranking

| No. | Risk Event | Chance | Effect | Risk Score |
|-----|------------|--------|--------|------------|
| 1 | Not receiving information from some departments | Low | Medium | Low |
| 2 | The designing company gets late in preparing the initial design | Medium | Medium | Medium |
| 3 | Not finding a suitable domain name for the site | Low | High | Medium |
| 4 | The construction company asks for more money than what is budgeted | Low | Low | Low |
| 5 | Receiving many changes to the design during the review of the initial design | Low | High | High |
| 6 | Testing takes more time due to the desire to collect many samples | Medium | Medium | Medium |

Template 18

## Opportunity Identification and Ranking

| No. | Opportunity Event | Chance | Effect | Opportunity Score |
|-----|-------------------|--------|--------|-------------------|
| 1 | One of the team members has experience in web design, so he can help greatly in preparing the technical requirements | Medium | Medium | Medium |
| 2 | The task of tendering can be reduced in duration if the company uses a specific list of preferred companies to design the site | Medium | Medium | Medium |
| 3 | We can start testing the general layout of the site before the site is fully finished | Medium | Low | Low |

Template 19

## The Risk Register

| No. | Risk | Risk Score | Control Option | Responsibility | Side Effect of the Option on the Project |
|-----|------|------------|----------------|----------------|------------------------------------------|
| 1 | Not receiving information from some departments | Low | Follow up with the different departments to collect the needed information | project team | - |
| 2 | The designing company gets late in preparing the initial design | Medium | Ask the designing company for a detailed time plan and closely monitor its progress | Bill | - |
| 3 | Not finding suitable domain name for the site | Medium | Start early to look for a domain name | Tony | - |
| 4 | The designing company asks for more money than what is budgeted | Low | In the selection, balance between both cost and quality | Bill | - |
| 5 | Receiving many changes to the design during the review of the initial design | High | Actively provide the GM, IT manager and Public Relations manager with information about the project and secure their approval of the scope | project team | - |
| 6 | Testing takes more time than planned | Medium | Team members can do testing simultaneously by each having a copy of the website on his/her personal laptop, thus collecting a large number of reviews | Bill + Tony | - |

Template 20

## Opportunity Register

| No. | Opportunity | Opportunity Score | Enhancement Option | Responsibility | Side Effect of the Option on the Project |
|---|---|---|---|---|---|
| 1 | One of the team members has experience in web design, so he can help greatly in preparing the technical requirements | Medium | The task of writing the tender to be given to that member | Tony | - |
| 2 | The task of tendering can be reduced in duration if the company uses a specific list of preferred companies to design the site | Medium | Coordinate with contracts department to use limited tendering for pre-specified companies | project team + contracts department | - |
| 3 | We can start testing the general layout of the site before the site is fully finished | Medium | As soon as the general layout is finished, testing can start with customers (in the showroom, for example) | project team | - |

# Example Two: A Family Trip to Paris

**Scenario:** Y ou have a spouse and two children, a 7-year-old girl and a 5-year-old boy. You have decided to go to Paris this summer for the purpose of relaxation and exploring France.

**Assumptions:**

* You will be managing a team of three people: your spouse, your daughter and your son.

* You intend to stay between one and two weeks.

* You live in the USA.

* You may make additional assumptions as necessary.

In this example, you should make accurate estimations of the budget and project task durations. To do so, you will need to do some research on the internet.

**Now, start filling out the templates!**

## Use This Page for Brainstorming and Notes

Template 1

# Non Financial Return on Investment

| Expected Return | Chance of Attaining the Return | Link with Organizational Goals | Cost |
|---|---|---|---|
|  |  |  |  |
|  |  |  |  |
|  |  |  |  |
|  |  |  |  |
|  |  |  |  |
|  |  |  |  |
|  |  |  |  |
|  |  |  |  |

# Use This Page for Brainstorming and Notes

Template 2

# Feasibility Study

| Description of the Idea |
| --- |
| |
| Available Alternatives |
| |
| Availability of Funding |
| |
| Can the Idea Be Implemented Technically? |
| |
| Availability of Human Resources to Run the Project |
| |
| Conflicts with Other Projects |
| |
| Recommendations |
| |

# Use This Page for Brainstorming and Notes

Template 3

# Project Charter

| | |
|---|---|
| Project Name | |
| Project ID | |
| Expected Duration | |
| Expected Budget | |
| Project Goals | |
| Project Owner | |
| Project Manager | |
| Start Date | |
| Signature | |

# Use This Page for Brainstorming and Notes

Template 4

# Project Team

| No. | Name | Tasks | Signa-ture |
|-----|------|-------|------------|
|     |      |       |            |
|     |      |       |            |
|     |      |       |            |
|     |      |       |            |
|     |      |       |            |
|     |      |       |            |
|     |      |       |            |
|     |      |       |            |
|     |      |       |            |

## Use This Page for Brainstorming and Notes

Template 5

# Lessons Learned Review

| No. | Previous Project Name | Lesson Learned | How to Use It |
|---|---|---|---|
|  |  |  |  |
|  |  |  |  |
|  |  |  |  |
|  |  |  |  |
|  |  |  |  |
|  |  |  |  |
|  |  |  |  |
|  |  |  |  |
|  |  |  |  |

## Use This Page for Brainstorming and Notes

Template 6

## Stakeholders Identification and Prioritization

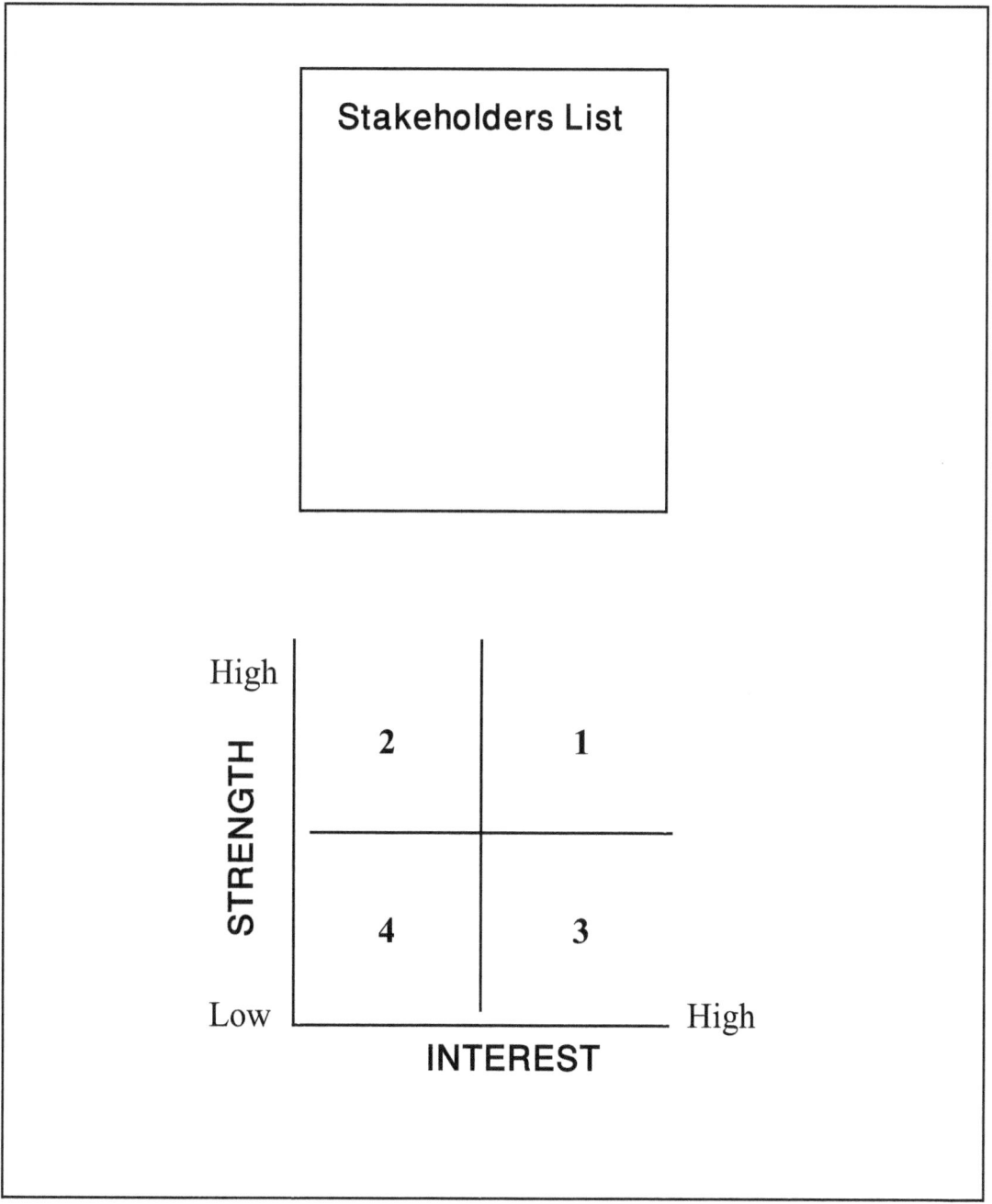

## Use This Page for Brainstorming and Notes

Template 7

# Stakeholders Requirements

| Stakeholders | Potential Requirements | How to Satisfy Requirements |
|---|---|---|
|  |  |  |
|  |  |  |
|  |  |  |
|  |  |  |
|  |  |  |
|  |  |  |
|  |  |  |
|  |  |  |
|  |  |  |

# Use This Page for Brainstorming and Notes

Template 8

## Information Sharing Plan

| Information | Receivers | Method of Sending | Frequency of Sending | Who Will Send the Information? |
|---|---|---|---|---|
| | | | | |
| | | | | |
| | | | | |
| | | | | |
| | | | | |
| | | | | |
| | | | | |
| | | | | |
| | | | | |

| Receiver | Contact | | Receiver | Contact |
|---|---|---|---|---|
| | | | | |
| | | | | |
| | | | | |

# Use This Page for Brainstorming and Notes

Template 9

# Project Scope

**Project Picture**

**What's Out**

**What's In**

**Use This Page for Brainstorming and Notes**

Template 10

## Project Map

Trip to Paris

## Use This Page for Brainstorming and Notes

Template 11

# Project Quality Requirements

| No. | Task | Required Quality | How to Measure It |
|-----|------|------------------|-------------------|
|     |      |                  |                   |
|     |      |                  |                   |
|     |      |                  |                   |
|     |      |                  |                   |
|     |      |                  |                   |
|     |      |                  |                   |
|     |      |                  |                   |
|     |      |                  |                   |
|     |      |                  |                   |

## Use This Page for Brainstorming and Notes

Template 12-A

**Project Map with Durations of Tasks**

## Trip to Paris

## Use This Page for Brainstorming and Notes

Template 12-B

## Network Diagram

```
┌─────────────┐                          ┌─────────────┐
│   START     │                          │    END      │
└─────────────┘                          └─────────────┘
```

# Use This Page for Brainstorming and Notes

Template 12

# Project Time Plan

**Time unit is ....**

| No. | Task | Time | 1 | 2 | 3 | 4 | 5 | 6 | 7 | 8 | 9 | 10 | 11 | 12 | 13 | 14 | 15 | 16 | 17 | 18 |
|-----|------|------|---|---|---|---|---|---|---|---|---|----|----|----|----|----|----|----|----|----|
|     |      |      |   |   |   |   |   |   |   |   |   |    |    |    |    |    |    |    |    |    |
|     |      |      |   |   |   |   |   |   |   |   |   |    |    |    |    |    |    |    |    |    |
|     |      |      |   |   |   |   |   |   |   |   |   |    |    |    |    |    |    |    |    |    |
|     |      |      |   |   |   |   |   |   |   |   |   |    |    |    |    |    |    |    |    |    |
|     |      |      |   |   |   |   |   |   |   |   |   |    |    |    |    |    |    |    |    |    |
|     |      |      |   |   |   |   |   |   |   |   |   |    |    |    |    |    |    |    |    |    |
|     |      |      |   |   |   |   |   |   |   |   |   |    |    |    |    |    |    |    |    |    |
|     |      |      |   |   |   |   |   |   |   |   |   |    |    |    |    |    |    |    |    |    |
|     |      |      |   |   |   |   |   |   |   |   |   |    |    |    |    |    |    |    |    |    |
|     |      |      |   |   |   |   |   |   |   |   |   |    |    |    |    |    |    |    |    |    |
|     |      |      |   |   |   |   |   |   |   |   |   |    |    |    |    |    |    |    |    |    |
|     |      |      |   |   |   |   |   |   |   |   |   |    |    |    |    |    |    |    |    |    |
|     |      |      |   |   |   |   |   |   |   |   |   |    |    |    |    |    |    |    |    |    |
|     |      |      |   |   |   |   |   |   |   |   |   |    |    |    |    |    |    |    |    |    |
|     |      |      |   |   |   |   |   |   |   |   |   |    |    |    |    |    |    |    |    |    |
|     |      |      |   |   |   |   |   |   |   |   |   |    |    |    |    |    |    |    |    |    |

# Use This Page for Brainstorming and Notes

Template 13

# Resources Table

| No. | Task | Human Resources | Cost/Comments | Equipment/ Materials | Cost/ Comments |
|---|---|---|---|---|---|
|  |  |  |  |  |  |
|  |  |  |  |  |  |
|  |  |  |  |  |  |
|  |  |  |  |  |  |
|  |  |  |  |  |  |
|  |  |  |  |  |  |
|  |  |  |  |  |  |
|  |  |  |  |  |  |
|  |  |  |  |  |  |
|  |  |  |  |  |  |
|  |  |  |  |  |  |
| Total |  |  |  |  |  |
| Project Budget |  |  |  |  |  |

**Use This Page for Brainstorming and Notes**

Template 14

## Technical Evaluation Template for Web Project

| Comparison Criteria | Weight | Company 1 | Company 2 | Company 3 | Company 4 | Company 5 |
|---|---|---|---|---|---|---|
| | | | | | | |
| | | | | | | |
| | | | | | | |
| | | | | | | |
| | | | | | | |
| | | | | | | |
| | | | | | | |
| | | | | | | |
| Total | | | | | | |
| Signature | | | | | | |

Template 15

# Make Your Own Change Request Template

Template 16

# Make Your Own Progress Reporting Template

# Use This Page for Brainstorming and Notes

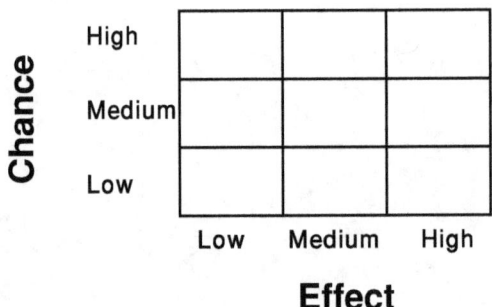

Template 17

# Risk Identification and Ranking

| No | Risk Event | Chance | Effect | Risk Score |
|---|---|---|---|---|
| | | | | |
| | | | | |
| | | | | |
| | | | | |
| | | | | |
| | | | | |
| | | | | |
| | | | | |
| | | | | |
| | | | | |

# Use This Page for Brainstorming and Notes

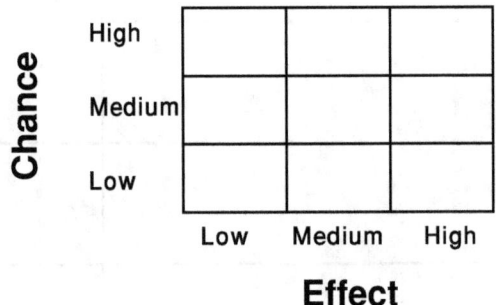

Template 18

# Opportunity Identification and Ranking

| No | Opportunity Event | Chance | Effect | Opportunity Score |
|----|-------------------|--------|--------|-------------------|
|    |                   |        |        |                   |
|    |                   |        |        |                   |
|    |                   |        |        |                   |
|    |                   |        |        |                   |
|    |                   |        |        |                   |
|    |                   |        |        |                   |
|    |                   |        |        |                   |
|    |                   |        |        |                   |
|    |                   |        |        |                   |

# Use This Page for Brainstorming and Notes

Template 19

# The Risk Register

| No. | Risk | Risk Score | Control Option | Responsibility | Side Effect of the Option on the Project |
|-----|------|-----------|----------------|----------------|------------------------------------------|
|  |  |  |  |  |  |
|  |  |  |  |  |  |
|  |  |  |  |  |  |
|  |  |  |  |  |  |
|  |  |  |  |  |  |
|  |  |  |  |  |  |
|  |  |  |  |  |  |
|  |  |  |  |  |  |
|  |  |  |  |  |  |

# Use This Page for Brainstorming and Notes

Template 20

# Opportunity Register

| No | Opportunity | Opportunity Score | Enhancement Option | Responsibility | Side Effects of the Option on the Project |
|----|-------------|-------------------|--------------------|----------------|-------------------------------------------|
|    |             |                   |                    |                |                                           |
|    |             |                   |                    |                |                                           |
|    |             |                   |                    |                |                                           |
|    |             |                   |                    |                |                                           |
|    |             |                   |                    |                |                                           |
|    |             |                   |                    |                |                                           |
|    |             |                   |                    |                |                                           |
|    |             |                   |                    |                |                                           |

# A Sample Project Book for a Family Trip to Paris

Template 1

## Non-Financial-Return on Investment

| Expected Return | Chance of Attaining the Return | Link with Organizational Goals | Cost |
|---|---|---|---|
| Spend quality family time | High | High | Money needed for the tickets, accommodations and expenses |
| Chance to see France | Medium | High | Father must take leave from work |
| Chance to try French food | High | High | - |

Template 2

## Feasibility Study

| Description of the Idea |
|---|
| Take the family on a 10-day trip to France |
| **Available Alternatives** |
| Go to Thailand or stay home |
| **Availability of Funding** |
| The budget needed will be around $7500 US, which is available |
| **Can the Idea Be Implemented Technically?** |
| Yes, flights are available to and from France. Accommodations can be booked via the internet using credit card. Visitors' visas can be obtained easily from the embassy. |
| **Availability of Human Resources to Run the Project** |
| Yes |
| **Conflicts with Other Projects** |
| The family doesn't have other activities planned during this period. Kids are on school holiday. |
| **Recommendations** |
| The idea was reviewed by the parents and kids and was approved to start planning for the trip |

Template 3

# Project Charter

| Project Name | Trip to Paris |
|---|---|
| Project ID | Trip 1 in 2010 |
| Expected Duration | 10 days |
| Expected Budget | $7500 US |
| Project Goals | To spend quality family time in Paris and to discover France |
| Project Owner | Entire family |
| Project Manager | Father |
| Start Date | 15-7-2010 |
| Signature | Father |

Template 4

# Project Team

| No. | Name | Tasks | Signature |
|---|---|---|---|
| 1 | Father | Project Manager. Responsible for completing the Project Book binder and all aspects of the project during the different stages | *father* |
| 2 | Mother | Help in completing the Project Book binder | *mother* |
| 3 | Daughter | Help in completing the Project Book binder | *daughter* |
| 4 | Son | Help in completing the Project Book binder | *son* |

Template 5

## Lessons Learned Review

| No. | Previous Project Name | Lesson Learned | How to Use It |
|---|---|---|---|
| 1 | Trip to Singapore | We booked cheep accommodations but they were far from shopping and attractions. We also faced trouble with transportation | Book accommodations close to tourist areas. |
| 2 | Trip to Singapore | Airline X was late in departure and service wasn't very good | Avoid this airline for this trip |
| 3 | Trip to South Africa | Many money-exchange services refused to exchange USD printed more than 3 years ago | Carry new dollar bills |

Template 6

## Stakeholders Identification and Prioritization

**Stakeholders List**

1 Father
2 Mother
3 Children
4 Mother in Law
5 Embassy
6 Travel Agencies

High

STRENGTH

Father
Mother
Children
Mother in Law
Embassy

Travel Agencies

Low     INTEREST     High

Template 7

## Stakeholders Requirement

| Stakeholders | Potential Requirements | How to Satisfy Requirements |
|---|---|---|
| Father | Visit the Louvre. Control spending | By visiting the Louvre and writing down daily spending |
| Mother | Good shopping and accommodations. Try new restaurants | By searching the travel sites for good accommodations in good locations |
| Children | Visit Euro Disney Land | Free one or two days to visit the theme park |
| Mother in Law | Know about the trip plan | By providing information about the trip |
| Embassy | Provide passport with full application before two weeks of the planned travel day | By complying with the requirement |
| Travel Agencies | Provide expected travel days in advance in order to find booking for the family | Provide travel dates in advance to make provisional booking |

Template 8

## Information Sharing Plan

| Information | Receivers | Method of Sending | Frequency of Sending | Who Will Send the Information? |
|---|---|---|---|---|
| Visa application with passport | French Embassy | In person | One time | Father + Mother |
| Information about hotels the family will stay in | Mother in Law | In person | One time | Mother |
| Dates of travel | Travel Agency | In person | As needed | Father |
| Receiver | Contact | | Receiver | Contact |
| French Embassy | Information Desk | | Travel Agency | Sonia at Tel. 123456789 |
| Mother in Law | Through Mother | | | |

Template 9

# Project Scope

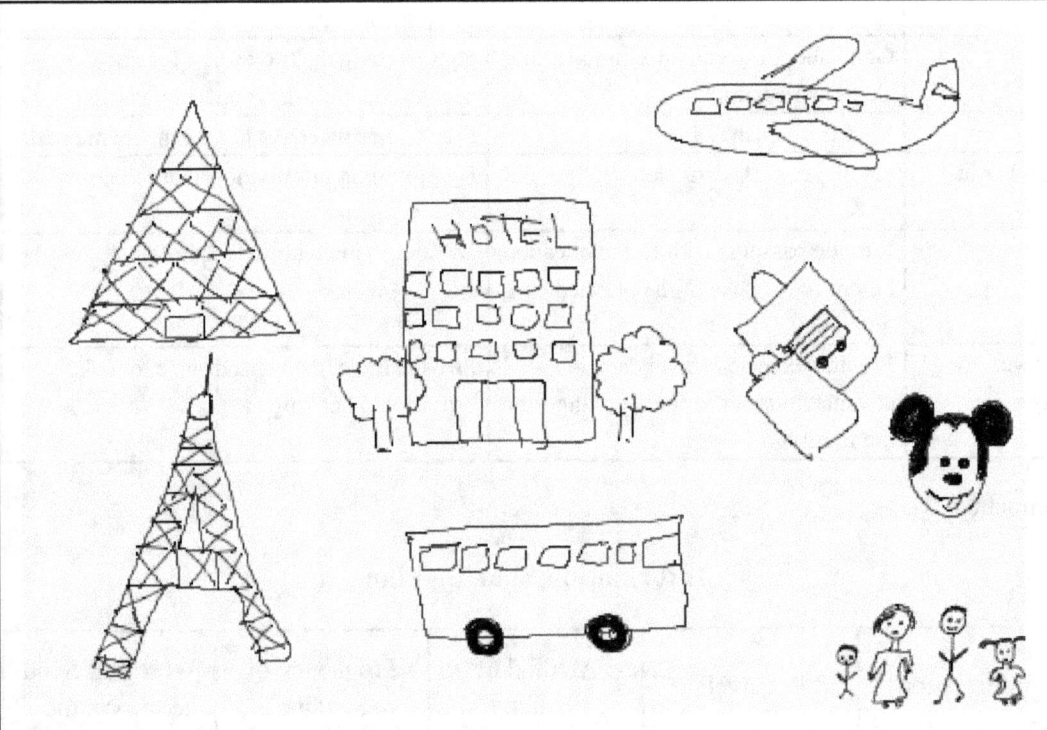

**What's Out**

- Visiting adjacent countries
- Business / first-class tickets
- Car rental
- 5-star hotels

**What's In**

- Airplane tickets
- Transfer to and from the airport
- Hotel accommodations
- City tour
- Visit to the Louvre
- Eating out
- Euro Disney
- Sightseeing
- Shopping

Template 10

# Project Map

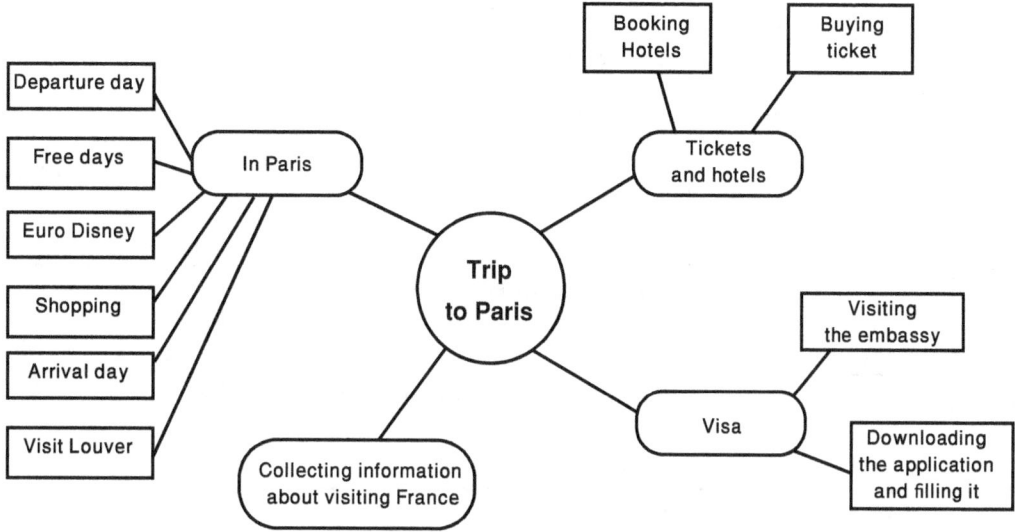

Template 11

## Project Quality Requirements

| No. | Task | Required Quality | How to Measure It |
|---|---|---|---|
| 1 | Collecting information about visiting France | Collect information from real people, not only from travel agencies. | Use travel forums and sites to search for the information |
| 2 | Downloading the application and filling it in | Completely fill in the visa application and sign it | No item should be left unfilled |
| 3 | Visiting the embassy | Come early to avoid the crowds and bring all required information | By leaving home at 7 in the morning heading to the embassy |
| 4 | Buying tickets | Flight to be direct and seats close together | - |
| 5 | Booking hotels | Hotels to be in central location close to attractions. To include breakfast and to be family-friendly. | - |
| 6 | Arrival day | Book a transfer to the hotel | - |
| 7 | Shopping | Buy gifts for mothers. Not to use credit card | - |
| 8 | Euro Disney | Arrive before opening time to avoid the crowds | Time of arrival |
| 9 | Visit Louvre | Arrive before opening time to avoid the crowds | Time of arrival |
| 10 | Free days | To include at least 2 free days. To make a city tour on one of these days | Number of free days and booking of a city tour |
| 11 | Departure day | Completely prepare the luggage. Arrive at airport 2 hours before departure | - |

Template 12-A

# Project Map with Durations of Tasks

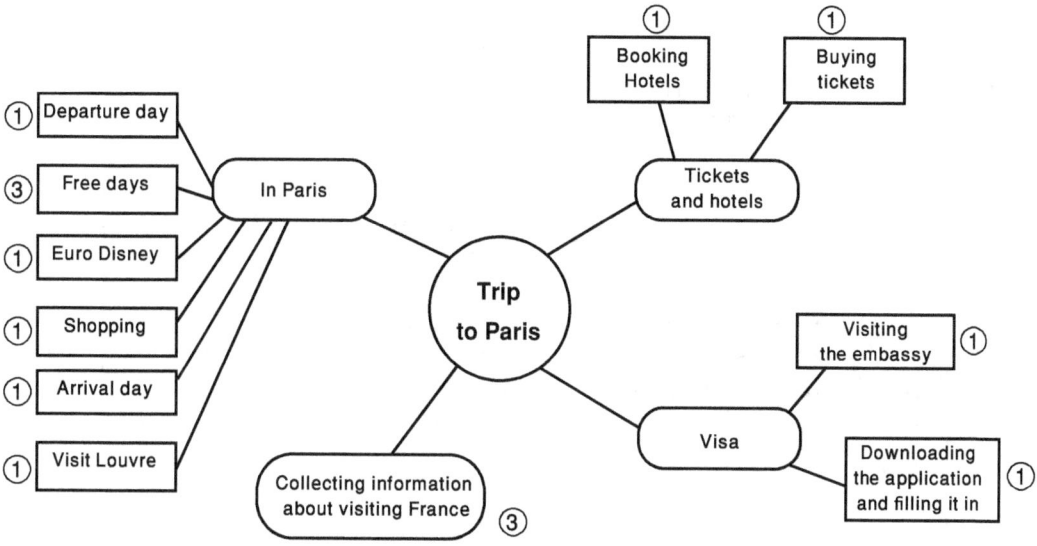

Template 12-B

## Network Diagram

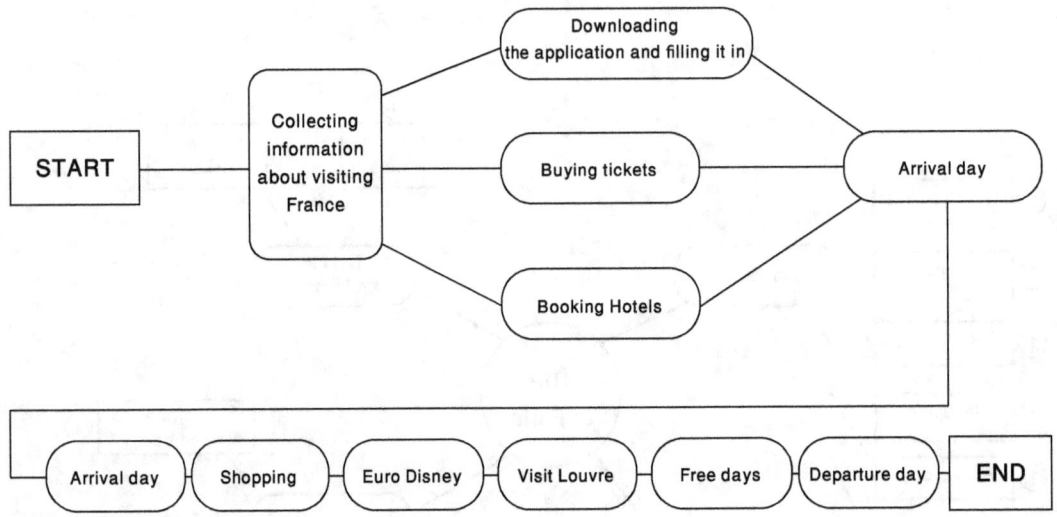

Template 12

## Gantt Chart

| No | Task | Duration | 1 | 1 | 1 | 1 | 1 | 1 | 1 | 1 | 1 | 1 | 1 | 1 | 1 | 1 | 1 |
|---|---|---|---|---|---|---|---|---|---|---|---|---|---|---|---|---|---|
| 1 | Collecting information about visiting France | 3 | █ | █ | █ | | | | | | | | | | | | |
| 2 | Downloading the application and filling it in | 1 | | | | | █ | | | | | | | | | | |
| 3 | Visiting the embassy | 1 | | | | | | █ | | | | | | | | | |
| 4 | Buying tickets | 1 | | | | | | | █ | | | | | | | | |
| 5 | Booking hotels | 1 | | | | | | | | █ | | | | | | | |
| 6 | Arrival day | 1 | | | | | | | | | █ | | | | | | |
| 7 | Shopping | 1 | | | | | | | | | | █ | | | | | |
| 8 | Euro Disney | 1 | | | | | | | | | | | █ | | | | |
| 9 | Visit Louvre | 1 | | | | | | | | | | | | █ | | | |
| 10 | Free days | 3 | | | | | | | | | | | | | █ | █ | |
| 11 | Departure day | 1 | | | | | | | | | | | | | | | █ |

Template 13

## Resources Table

| No. | Task | Human Resources | Cost / Comments | Equipment / Materials | Cost / Comments |
|---|---|---|---|---|---|
| 1 | Collecting information about visiting France | Father and mother | - | PC | - |
| 2 | Downloading the application and filling it in | Father and mother | - | PC | - |
| 3 | Visiting the embassy | Father and mother | - | Car + Visa fee for family | $300 |
| 4 | Buying tickets | Father | - | Cost of tickets | $1750 |

| 5 | Booking hotels | Mother | - | Cost of hotel transfer | $100 |
| 6 | Arrival day | Family | - | - | - |
| 7 | Shopping | Family | - | - | $750 |
| 8 | Euro Disney | Family | - | - | $300 |
| 9 | Visit Louvre | Family | - | - | $200 |
| 10 | Free days | Family | - | - | $500 |
| 11 | Departure day | Family | - | Cost of hotel transfer | $100 |
| Total | $5370 US | | | | |

Template 14

## Selection Template

For this project we will have two evaluation forms: one for selecting the airline and one for selecting the hotel.

**First:** Airline selection criteria

| Comparison Criteria | Weight | Airline 1 | Airline 2 | Airline 3 | Airline 4 | Airline 5 |
|---|---|---|---|---|---|---|
| Direct flight | 25 | | | | | |
| Suitability of flight timing | 15 | | | | | |
| Reputation of good service | 5 | | | | | |
| Luggage allowance | 5 | | | | | |
| Cost | 50 | | | | | |
| Total | 100 | | | | | |
| Signature | | | | | | |

**Second:** Hotel selection criteria

| Comparison Criteria | Weight | Hotel 1 | Hotel 2 | Hotel 3 | Hotel 4 | Hotel 5 |
|---|---|---|---|---|---|---|
| Location | 30 | | | | | |
| Breakfast included | 5 | | | | | |
| Family-friendly | 5 | | | | | |
| Easy access to metro | 10 | | | | | |
| Cost | 50 | | | | | |
| Total | 100 | | | | | |
| Signature | | | | | | |

Template 17

## Risk Identification and Ranking

| No. | Risk Event | Chance | Effect | Risk Score |
|---|---|---|---|---|
| 1 | Not finding the transfer service to the hotel | Low | Medium | Medium |
| 2 | Losing one or more of the suitcases | Low | Medium | Medium |
| 3 | Money gets stolen | Low | High | Medium |
| 4 | The Louver is closed on the day of visit | Low | High | Medium |
| 5 | One of the kids gets sick | Low | High | Medium |
| 6 | The flight back home is delayed | Low | Medium | Medium |
| 7 | Strike in Disney park | Low | Medium | Medium |

Template 18

## Opportunity Identification and Ranking

| No. | Opportunity Event | Chance | Effect | Opportunity Score |
|---|---|---|---|---|
| 1 | Can visit the Louvre and do a city tour on the same day and, thus, can visit Euro Disney for two days | High | Medium | High |
| 2 | Can get a discount on the Louvre if tickets are booked through the internet | High | Medium | High |

Template 19

## The Risk Register

| No. | Risk | Risk Score | Control Option | Responsibility | Side Effect of the Option on the Project |
|-----|------|------------|----------------|----------------|------------------------------------------|
| 1 | Not finding the transfer service to the hotel | Medium | Print out the directions to the hotel to be able to take a taxi | Mother | - |
| 2 | Losing one or more of the suitcases | Medium | Distribute clothes among all suitcases | Mother | - |
| 3 | Money gets stolen | Medium | Distribute money in different places. Use the hotel safe. Bring credit card for emergency | Father + Mother | - |
| 4 | The Louvre is closed on the day of visit | Medium | Change the plan and make that day a free day | Father | - |
| 5 | One of the kids gets sick | Medium | Take medicine. Eat in good restaurants. Avoid crowded places as possible | Mother + Kids | - |
| 6 | The flight back home is delayed | Medium | Book one more night in hotel | Father | Cost of additional night's stay |
| 7 | Strike in Disney park | Medium | Change the plan and go to the park on another day | Family | - |

Template 20

## Opportunity Register

| No | Opportunity | Opportunity Score | Enhancement Option | Responsibility | Side Effect of the Option on the Project |
|---|---|---|---|---|---|
| 1 | Can visit the Louvre and do a city tour on the same day and, thus, can visit Euro Disney for two days | High | Go to the Louvre early | Family | - |
| 2 | Can get a discount on the Louvre if tickets are bought through the internet | High | Buy tickets through the internet | Mother | - |

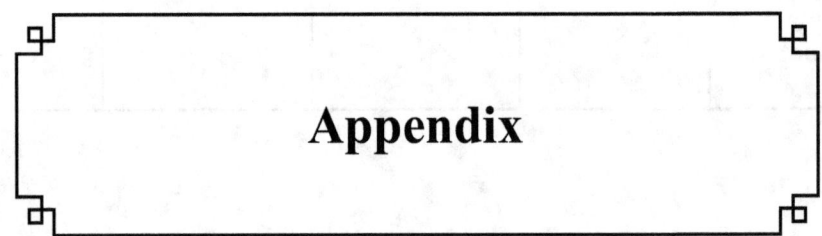

# Appendix

## The Project Simulation Game:

Simplifying Project Management Through Concrete Learning

and the Use of the "Project Book"

Abdulla J. Alkuwaiti

### The Problem

Since 2008, a municipality in the Middle East has adopted a "management by projects" approach to all its activities and linked all its projects to a strategic plan. The strategic plan contains more than 200 projects ranging from US$50,000 to more than US$200 million. The difficulty is with the project managers, mostly fresh graduates with no or little experience in project management. They jump directly to project execution and provide excuses for bypassing the project planning stage such as "Project management is a dry subject and very difficult to learn," and «Planning is a waste of time and its benefits are minimal in the practical word." The challenge is to present project management (as a science) in an innovative way that grasps the attention of those project managers.

### The Method

The project management office (PMO) started investigating ways to make the project managers practice and implement methodologies and techniques of project management. It was clear that training was needed, but records showed that different project management courses were offered yearly by the training department. After analysis, it was concluded that the way training was conducted failed to transfer the knowledge gained of project management from training manuals to the actual projects, for the following reasons:

- Failure to address the uniqueness of the municipality. Project managers expressed frustration that what is taught in training courses seldom represents what they face in day-to-day work. For example, most of the projects in the municipality are done via a consultant and a contractor, whereas training courses only teach about situations in which the project manager is the one who will design and implement the project.

- The way that exercises were presented. It was found that few exercises were discussed in training courses, and that the way that they were presented lacked the creativity to catch the project manager's attention. (As one project manager put it, "They give you three to four pages of a single-spaced case study of a project without even an illustration and expect you to relate to it!")

- The amount of information. Project managers complained that the amount of information presented to them was overwhelming and made them think of project management as complex and difficult.

To improve training, the PMO conducted extensive research to identify new training methods and decided to provide a training course that is centered on a game that simulated a real-life project. The use of the game served the following purposes:

- Focusing on concrete vs. abstract learning. A model of the project was constructed using toys. In this way, the participant can more easily relate to the case study by interacting with the toys directly.

- Fun and simplicity. The way that the game was designed, using toys, provided a fun atmosphere to the course in general. In addition, training manuals were used as a minimum to avoid any sign of complexity.

- Competition. Participants were divided into teams to provoke competition. This also enabled the lecturer to use group pressure in a good way, where being part of a team made members keen on attending and focusing on the different lectures so that they wouldn't let their team down.

- Purposely designed plot and using templates as checkpoints. A great deal of thought was put into selecting the game so that participants could relate to it in their day-to-day work. The game was designed in a way that enabled the lecturer to go over different elements of project management methodologies in a smooth and logical manner. In addition, the use of templates made it easy for participants to know when they had finished a particular task (which was extremely helpful in discussing the different elements of the project planning stage)

**The Training**

The course lasted for 2 days with 17 participants and was started with a brief introduction about project management followed by an introduction to "brainstorming" and how to conduct it. After that the game was started with a problem statement and teams were created. The case study was about the recent changes in weather conditions in the area which had caused some hurricanes and floods, and the project (that is, the game) was to build a weather station tower.

The remainder of the course followed the model below, where a brief description of a project management subject was given, followed by observing the game model and filling the template related to the discussed subject and, finally, discussing the real-life benefit of the subject.

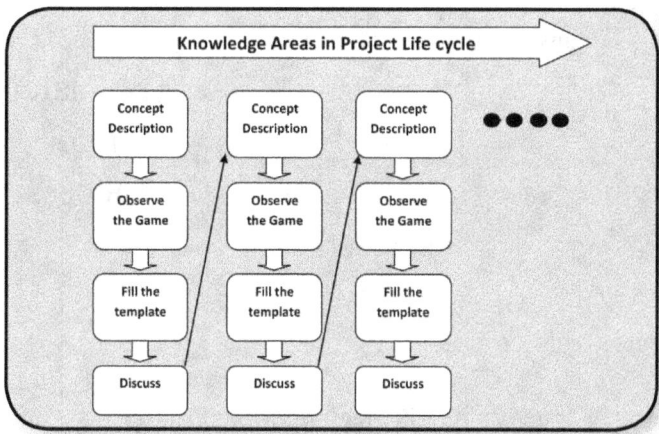

**Figure 1: The model of the training course.**

The topics discussed were the ones that the PMO felt project managers most needed to know, such as scope management, WBS, and time and risk planning. To make the course dynamic and avoid boredom, each topic was discussed on average for just 1.5 hours.

**The Templates and the Project Book**

A big emphasis in the course was to fill templates. Templates were designed with the objective that no template would be more than one page long. In addition, templates were arranged in a binder called the "Project Book". The Project Book had sections for each part of the project life cycle (from initiation to closeout) with relevant templates placed in each section. The "Project Book" served a very important role; it gave the project managers a sense of the importance of the project planning stage (as it contained the majority of templates). In addition, once a template was completed it signaled the completion of the subject under discussion and "gave permission" to move to the next one. It should be noted that although working in teams, every participant was asked to fill out his or her own template after conducting a brainstorming with the team, and thus every participant should have had a completed "Project Book."

The use of templates was found to be very important because it triggered brainstorming, which in turn triggered communication with people interested in the project. Also with each template completed, a better understanding for the project was developed. The model shown in Figure 2 explains the benefits of using the templates:

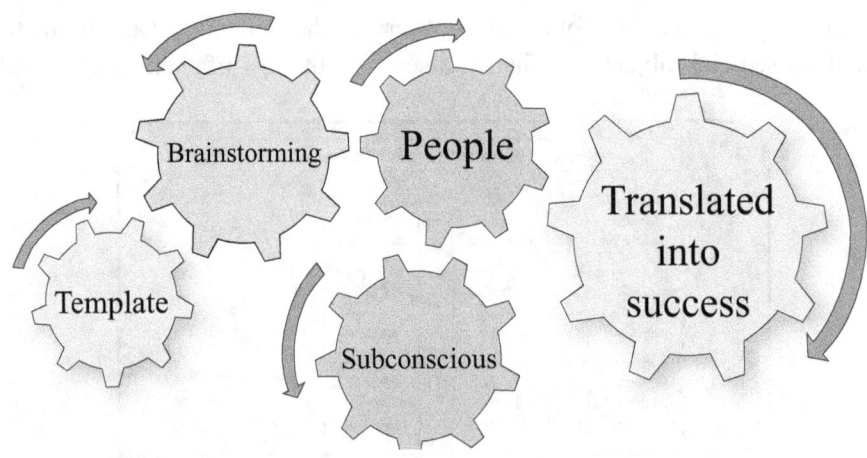

**Figure 2: The Project Template Model**

**Observation and Recommendations**

1 As expected, using a physical model for the case study proved to be very helpful. It improved brainstorming because participants could imagine what the project would look like by seeing a physical model of the construction site and the tower). It also reduced the abstractness of the game which reduced conflicts and disputes usually generated when different people interpret a written exercise.

2 When designing a game or exercise, be sure of the plot. I remember attending one training course that was also centered on a case study; however, the case study was displayed only on paper and the game did not flow smoothly from one project stage to another. Conversely, in this training, deep thinking was put to connect the different project management Knowledge Areas to the game so that maximum benefit can be attained (for example, during execution phase, the participants were handed three proposals from imaginary contractors to construct the weather tower, so that they can implement the selection criteria they defined during project planning phase).

3  Participants liked the templates and found them very useful. This can be attributed to several factors. First, templates were designed to fit in only one page with the minimal information possible. Second, templates were unique in that they provoked the user to try seeing things in new way; for example, the template to develop the project scope required the participant to draw a picture of the scope. Finally, templates were numbered in sequence, which provided a sense of order on how to proceed about filling them (for example, there were two templates for the project time plan, one to be filled after completing the WBS and then edited and finalized in another template after the risk management template was completed, to stress the fact that risk management effect the time baseline).

4  Don't be afraid of simplicity. In the training course, we tried to simplify project management concepts so much that we were afraid that participant would be annoyed by the "oversimplification." What we discovered, however, was that many participant were being introduced to the concepts for the first time and asked for more explanation (for example, on topics such as stakeholder management and even the WBS!). This might be attributed to what is sometimes called "the curse of knowledge," which describes the situation in which people become so knowledgeable about a particular subject that they forget how difficult it is to learn it initially.

5  The use of "The Project Book" gave a concrete meaning to the different stages of the project life cycle. This was especially important to distinguish the initiation and planning stages as important stages worthy of the project manager's attention (remember in the problem statement, project managers in the municipality were bypassing project planning and jumping directly to execution by writing the RFP, the request for proposals to potential consultants). During the course, participants were not allowed to start execution until they filled all the templates in the planning section of the project book.

What's Next?

Explore the concept of "The Project Book" further to see if project managers will actually use it in their real projects and what improvement on it should be made. In this paper it was shown that the use of a structured and simple framework (made of templates) has the potentials of improving project planning which will defiantly have a positive impact on projects progress. In addition, the paper implies that new methods for training should be tried to increase the awareness of project managers about project management.

# index

# index

www.ingramcontent.com/pod-product-compliance
Lightning Source LLC
Chambersburg PA
CBHW081111170526
45165CB00008B/2407